Turf Managers' Handbook for Golf Course Construction, Renovation, and Grow-In

Bud White

WILEY

JOHN WILEY & SONS, INC.

Linoprint Composition Co., Inc.

180 Varick St., New York, NY 10014
Phone: (212) 675-2945 Fax: (212) 645-9257 E-Mail: info@linoprint.com

Linoprint Job Number: **70334**
Customer Name: **John Wiley & Sons**
Attention: **Julie Castaldo**
Customer Job Title: **C0790/Turf Manager's Handbook**

OPRT. CH/CUSTOMERS A/WILEY/70334
Date of First Setting: **12-11-02**
Date of Third Revise: **12-20-02**
Typefaces: **Times, Gill Sans**

Trademark Notice: Product or corporate names may be trademarks or registered trademarks, and are used for only identification and explanation, without intent to infringe.

This book is printed on acid-free paper. ♾

Copyright © 2000 by John Wiley & Sons, Inc. All rights reserved.

Published by John Wiley & Sons, Inc., Hoboken, New Jersey
Published simultaneously in Canada.

For general information on our other products and services or for technical support, please contact our Customer Care Department within the United States at (800) 762-2974, outside the United States at (317) 572-3993 or fax (317) 572-4002.

Wiley also publishes its books in a variety of electronic formats. Some content that appears in print may not be available in electronic books. For more information about Wiley products, visit our web site at www.wiley.com.

Library of Congress Cataloging-in-Publication Data:

White, Charles B.
  Turf managers' handbook for golf course construction, renovation and grow-in /
by Charles B. White
    p.   cm.
  Includes bibliographical references and index.
  ISBN 1-57504-110-3
    Golf courses—Design and construction.   2. Golf courses—Maintenance.
    I. Title.

GV975.W45 2000
796.352'06'8—dc21

99-087983

Printed in the United States of America

10 9 8 7 6 5 4 3 2

To Karen and Britt,
God's special blessing to me.

ACKNOWLEDGMENTS

I would like to express my appreciation to Rod McWhirter; Rick Elyea; Brad Kocher; Beall, Gonnsen & Company; and Mac Smith for photo/technical assistance; Dr. Charles Peacock for content editing; and special thanks to Steve Batten for all the artistic drawings. I'm especially grateful to Lisa Alfonzo for transcription and editing. I could not have done the job without her.

Finally, to Karen, my wife and best friend, my most sincere appreciation. She has been supporter and helper through every aspect of this book. She loves this industry as I do. She is the very best!

PREFACE

My turfgrass career began in 1970 at the age of fifteen. My first job was hand-raking bunkers and night watering. Central control pop-ups or mechanical rakes were yet to be a factor in turfgrass management.

I developed a love for the game of golf as a young child and, after working two summers on the golf course, I knew it would be my life's profession. Following high school I attended a two-year school—Catawba Valley Technical College—for a turf management degree. My professor, Harry DuBose, persuaded me to go on for a B.S. degree, so I transferred to and earned my degree from Tennessee Tech University before continuing my education and earning an M.S. at Clemson University. Needless to say, Harry had a tremendous impact on my education.

As I finished my M.S. degree I was interviewed by Monty Moncrief of the USGA Green Section for the Southeastern Regional Agronomist position. My youth was offset by a good balance of golf course maintenance, construction, and tournament experience. I learned more from Monty than I could ever realize. He was one of the greatest men our industry has ever known.

During my tenure with the Green Section I witnessed and participated in the incredible construction and grow-in activity which soon became my first love of the industry. To date I have consulted at over 100 golf course projects in construction, renovation or grow-in and have been directly involved in the construction and grow-in of four courses. The majority of my consulting work in these areas includes quality control, finish construction, establishment, maintaining a grow-in program, and preparing for opening.

Golf course development covers a wide array of jobs from

initial conceptual drawings to opening day, and can be a complicated and multifaceted process that requires much detailed planning. Centered in all these processes is finish construction and grow-in of the golf course. Much has been published on golf course construction, architecture, golf course building and renovation, and recommendations for opening; however, there is a window during this entire process, known as grow-in, which seems never to be adequately addressed.

The premise of this book stems from a Golf Course Superintendents Association of America (GCSAA) Seminar, "The Superintendent as Grow-in Manager." This expanded seminar has been translated here in an effort to give the reader an "in-the-field," "walk-through" guide. The discussion of final construction and grow-in management in this book is designed to be practical and hands-on. Furthermore, this book details the grow-in process and its many duties and priorities as it progresses. Discussions include not only the "how-to's," "what if's," and "what now's," but also agronomic theories in the areas of:

- Irrigation management
- Fertility management
- Mowing philosophies
- Erosion and sediment control

A balance of field experience and agronomic theory is necessary to fully understand the concepts involved with proper grow-in management and to anticipate and troubleshoot on-site throughout the management process. Combining these two philosophies produces an in-depth understanding of quality control during finish construction and grow-in programming during the establishment phase.

Purpose

As golf course construction ends and planting begins, the golf course is actually only 70% to 75% complete. The grow-in or establishment phase is a critical time in the development process because the creation of a turf cover (playing surfaces) brings to bear the full beauty of the golf course and its design in relation to the surrounding landscapes. A poor grow-in can ruin many design features through erosion; in extreme cases, design features are lost indefinitely without major renovation. The grow-in process is neither simple nor inexpensive, but it is manageable and reasonable in cost as long as a good understanding of the process and requirements for hands-on field management exists.

There is no detailed reference available today pointing out finish construction concerns from an agronomic standpoint, or how a turf manager can monitor finish construction activities to determine agronomic performance. Neither is there a source for detailed grow-in information on how the many job aspects of the grow-in period are coordinated. A major goal of this handbook is to outline and detail the many quality control issues concerning an owner representative, and to serve as a guide for questions that should be asked after construction (e.g., requesting copies of the greens mix and tee sand specs).

Often an owner's false perception of grow-in is that it is an inexpensive procedure that is relatively short in duration. Contrary to popular belief, construction cost surveys today do not include the cost of grow-in, and yet the grow-in expense must be absorbed before opening the course—the same as greens construction or irrigation purchase and installation. This flaw in thinking can be costly, as overlooked and underfunded grow-in details can result in future problematic course conditions. This is especially true since grow-in occurs when the significant expenditures of design and construction have recently been completed.

This handbook will stress the magnitude and importance of finish construction and grow-in as it relates to turf maintenance, overall turf performance and playability of the course from its opening into its future. It will address the concerns and perspectives mentioned above and illustrate how all job duties and unexpected surprises are coordinated through the grow-in process.

Objectives

The objectives of this handbook are to:

- Promote a better understanding of the critical phase of finish construction and grow-in from the turf manager's and the owner's or developer's perspective.
- Provide guidelines for quality control monitoring as the owner representative.
- Provide guidelines for finish construction details as they relate to agronomic concerns and playability.
- Demonstrate the practical management of key job duties and how they are prioritized as the grow-in progresses.
- Show how to anticipate unexpected surprises and how to manage and correct them.
- Develop accurate agronomic programs for the grow-in process.
- Enable the turf manager to accurately develop a critical path and budget for the project from finish construction through grow-in to opening.
- Provide a resource for superintendents to educate owners or developers on the details of the grow-in sequence and its budgetary requirements.
- Provide "common sense" guidelines for managing and troubleshooting the grow-in process to allow superintendents to predict and anticipate.

The references in the Reference list were used to develop the most accurate program recommendations possible and to provide a complete reference source by subject. Because small amounts of information were compiled from many sources and would be impossible to identify by footnote, the reference list is included at the end of the book.

The illustrations and examples contained herein are designed for educational purposes only. There is no intention of commercial preference, and any illustration that includes commercial indications should not be interpreted as an endorsement or recommendation of a manufacturer or product.

CONTENTS

APPENDIX

CHAPTER ONE

WHERE TO START?: THE BIG PICTURE

Project Development

A vital role of the golf course superintendent during course construction and development is to have a solid understanding of the project development. This "big picture" entails two large areas of planning and emphasis:

1. The initial critical path between construction to planting, and
2. Setting a realistic opening date for providing the necessary quality completion.

The critical path found in Appendix 1 is a path that would apply for a golf course in the transition zone of the eastern United States with bermudagrass tees, fairways, and intermediate roughs; bentgrass putting greens and collars; and turf-type tall fescue in the deep roughs. This critical path details the various jobs that must take place through fall, winter, and spring from the time greens and deep roughs are planted until opening day the following spring. A critical path can serve as an excellent flowchart that demonstrates to property management personnel the amount of work needed during the fall, winter, and spring months for a successful grow-in. The critical path also clarifies the necessary tasks that exist even after primary grow-in is complete.

Developing a critical path also offers an excellent means of justifying why a premature opening date is so costly. Owners, developers, and even project managers not well versed in golf course development may expect an opening date early in the spring, which is usually eight or nine months after construction completion. What is oftentimes forgotten is the four or five winter months, with little or no turf growth in most environments, that must be endured before spring arrives. The length of time from construction to opening day must be considered in conjunction with whether it is growing or nongrowing season; in other words, the time of year in relation to planting date and climate is what affects an opening date. As a general rule, a cool season golf course with a late summer or fall planting will have a midsummer opening. This allows substantial maturity and density achievement through early summer.

If you are developing a critical path for a golf course project, be sure to study the example in Appendix 1 and outline a new plan carefully tailored to your situation and in greater detail to help management understand all the process involved from final construction details until opening. All too often golf courses open too early and actually discourage players with poor playing conditions, especially "around the edges." "This golf course will be great when the grass is covered . . . we'll try it again in six or eight months" can kill the needed cash flow of a new course—a fate that can easily be avoided by delaying opening day. This "extra time" also allows significant time to complete the many details involved in the final stages of grow-in. These details should also be carefully mapped out for golf course management.

Figure 1.1. *Final construction has many details to coordinate.*

Golf Course Superintendent/Owner Rep

Hiring a golf course superintendent early in the process is essential for the successful coordination of all project management personnel as course construction and grow-in evolves (Figure 1.1). A superintendent is the best qualified to serve as an owner representative, or "rep." I strongly recommend that weekly or biweekly meetings between management personnel and the superintendent be conducted to keep everyone informed of development progress and to address potential or developing problems. These organized "production" meetings would address issues such as:

1. Rain delays with construction;
2. Unearthing of rock and subsequent blasting that is necessary, as public notice may need to be given;
3. Coordination of crossing project utilities on the golf course, and
4. Coordination of rest station construction and development on the golf course during course construction.

3

The owner rep should also help synchronize the golf course development versus the surrounding project development, if any. Coordination is critical; for example, an irrigation system could be significantly damaged by water or sewer line installation on the project if "the right hand does not know what the left hand is doing." Such concerns to be addressed include:

1. Road crossings for cart paths and irrigation on the golf course coordinated with utility construction along road shoulders (e.g., sewer, water, electricity),
2. Crossing the golf course with project utilities or storm drainage,
3. Coordination of well-marked back survey lines for lots adjacent to golf course, and
4. Setting guidelines regarding real estate sales personnel traffic on the golf course itself, as unnecessary and damaging vehicular traffic can occur when real estate sales people show surrounding golf course property.

The owner rep should also consider early planning for the construction of the golf course's rest station, as permits must be acquired for rain shelters, rest rooms, or a halfway house. The necessary arrangements must also be made to supply water, power, and sewer services to these structures, as well as lightning protection. The needs of any adequate rest station must be planned during the construction phase to allow coordination with golf course construction. Again, damage to golf course irrigation, drainage, or even cart paths can result from the poorly-timed installation of a rest station's septic tank, for example.

The owner rep's management team can make the difference between an arduous, stressful construction and grow-in and a cooperative, team effort approach that makes these busy months less harried. Common members of the management team at a minimum include the (1) superintendent, (2) project manager, (3) director of golf, (4) real estate sales director, and (5) owner or developer. All key personnel must have a complete grasp of the project flow, as each member can best manage their respective areas of expertise. The management team is a decision-making body for setting specifications such as grassing type, opening date, golf course amenities, and desired maintenance levels. The team also provides the most efficient means of working with the golf course architect and builder.

The owner rep is primarily responsible for the coordination and direction of the management team. A proactive approach to regular meetings and regular golf course walk-throughs with the architect, builder, and team members will ensure a smooth and efficient progression of construction and course readiness. Walk-throughs are very productive when all members are present, and a typed, preplanned agenda should be distributed before each walk-through occurs. Having the management team actively participate in walk-throughs gives them a better knowledge of the golf course from a design standpoint and an understanding of why certain construction activities were accomplished to meet the design goals from playability, drainage, and maintenance aspects or even aesthetics (Figure 1.2).

Figure 1.2. The management team must have routine walk-throughs.

A final management team walk-through of the entire golf course should also be scheduled at the end of grow-in. The walk-through can even coincide with the architect's follow-up visit to help the superintendent establish fairway mowing contours and other final design setups.

A walk-through by the management team with the architect at this time allows for open discussion about any concerns or adjustments that may need to be made now that the course is complete and issues are more clearly visible. Modifications that might be addressed include changing an established fairway contour mowing pattern, a final discussion of tree placement that may deviate from the master plan, or the relation of a green or tee complex to a surveyed lot line that had not been previously identified. In short, the walk-through can serve as one of the most beneficial and productive team meetings of the entire development phase.

Job Duties

Superintendents can provide invaluable expertise during the planning and construction phases of the golf course. Such duties may include:

1. Providing input to proper irrigation system specs and needs for the golf course (e.g., controller locations, quality control, assessing coverage).
2. Assisting with drainage concerns on the course as it would relate to maintenance (specifically, catch basin locations).
3. Identifying the responsibility of erosion and sediment control during and after construction and becoming familiar with the erosion and sediment plan.
4. Developing the best root zone mix for greens and tees—through representative sampling.

5. Certifying the golf course in the Audubon Cooperative Sanctuary Program.
6. Providing input to grass selection and slope steepness as it relates to maintenance costs and operations.
7. Providing budget guidelines for grow-in, the first year's maintenance, and projected normal maintenance costs.
8. Providing an equipment list, maintenance facility design considerations, and stocking lists.
9. Installing cart paths and bridges. This includes reviewing locations, widths, elevations, and materials with architect or owner.
10. Establishing a tree nursery for replacement and additional planting on the golf course according to the master plan.
11. Monitoring continued testing of all root zone materials on a regular basis for conformity.
12. Collecting soil samples for all preplant soil testing needs to ensure proper materials and rates, and coordinating these with various soil type needs.
13. Working closely with sales and marketing personnel during the final construction and grow-in phases to coordinate proper development sales traffic on and around the golf course.
14. Reviewing request for payments (RFP) submitted for accuracy and progress.
15. Making sure asbuilt drawings are properly developed and accurately recorded.

Organization and Record Keeping

Undoubtedly the most important job duty for the superintendent serving as an owner rep during construction is coordinating regulatory and compliance issues along with the agronomic considerations of the project, but ensuring a high quality asbuilt drawing preparation is also critical. Asbuilts are an integral part of record keeping for the golf course and will be relied on heavily in the future to quantify specific areas of construction. Golf courses are large, multiacre affairs, and construction details of buried items will quickly be forgotten if left to memory. Details may include:

1. Irrigation layouts and measurements of key structures;
2. Layouts of all storm drainage;
3. Information about all tile work, including green and bunker tile system design and outfall channeling;
4. Locations of all catch basins;
5. Locations of bury pits;
6. Easements for all storm drain and services for development that cross the golf course, including sewer, water, telephone, power, or cable routings;
7. Service routings for rest stations; and
8. Notations on specific site problems (e.g., shallow underlying rock).

Such details are recorded on a set of golf course plans and the green complex detail booklet. This becomes a daily "field notebook" for record keeping. During construction it is advisable to make note of areas where future problems could occur as the course ages. For example, a wet area may be cause for future concern. Some wet areas are addressed through the construction process by grading; however, these areas may resurface as wet springs that will need to be drained with a tile drain. Bury pits must also be identified in your field notes so their locations can be included on the asbuilt drawings. The location and construction of bury pits are key: they should never be at level grade because, even if constructed properly, they will settle over time; and they should never be in contact with irrigation lines, green or tee complexes, drainage or cart path installations. Figure 1.3 illustrates the result when their location is not accurately mapped. Recognizing potential problems will allow you to better identify and budget for them in the future.

Figure 1.3. Bury pit locations are critical.

Field notes should not only identify and locate items but include specifications on those items: for example, the depth of fill cover over a bury pit; the depth and location of large storm drainage; and the location and depth of development service easements across the course including sewer, water, power, gas, or cable. Green complex detail plans are also helpful with making specific notes. Noting a potential air movement problem on the greens detail upon completed construction should precipitate discussion and resolution of the problem between the owner rep, contractor, and owner.

A daily, dated logbook should be maintained during construction to document activities, progress, and problems. Daily weather data is also included to record temperatures and rain data for future reference. This is important for erosion and sediment control reporting when all rain events are detailed.

Other logbook entries include quality control (QC) notes (kept in conjunction with field plan notes), erosion and sediment control monitoring (also in conjunction with field plans), and other pertinent notes and "To Do" lists which are routinely prioritized for efficiency (Figure 1.4).

Figure 1.4. Erosion and sediment protection around wetlands is paramount.

Throughout this book there are many examples of good technique and management as well as small and major mistakes. These serve to show the record-keeping power of the camera. A project superintendent should fully utilize a camera during any construction/renovation project to detail activities, materials, and technique.

Pictures can also serve as a positive tool in identifying the "before and after" of renovation work and serve as an educational tool for green committee or board presentations. A combination of slides (for presentations) and prints is probably best. However, digital cameras offer the best of both worlds through computer generation of pictures (Figures 1.5–1.8).

Once golf course construction is complete, the golf course contractor should provide a detailed set of asbuilts identifying all storm drainage; tile drainage; locations of tees, greens, and bunkers in relation to the hole corridor; cart paths; tree lines; and bury pits. The irrigation asbuilt is updated after installation and should be copied onto a Mylar overlay sheet to the same scale as the golf course asbuilt supplied by the contractor. When the Mylar overlay is placed upon the golf course asbuilt, a complete underground

Figure 1.5. *Course renovation—before.*

Figure 1.6. *Course renovation—after.*

11

Figure 1.7. *New construction—before.*

Figure 1.8. *New construction—after.*

picture of all aspects of subsurface construction on the course is presented.

The golf course contractor is not responsible for including any project development service easement that may have crossed the golf course during construction; rather, it is the owner rep's responsibility to include these on the asbuilt and provide the contractor with measurements from fixed points. Appendices 2 and 3 show examples of field notes made on the green detail of a construction plan and properly drawn asbuilts. These examples plainly show how important field notes and "accurate measurements" are to future activities and problem solving on the golf course. This is especially true as the real estate development phase around and within a golf course is constructed.

Request for payments (RFPs) are usually made monthly by the golf course contractor to the owner. It is the owner rep's responsibility to review these RFPs to determine accuracy of work completed versus submitted work. This is usually recorded as the percent completed of each line item of the construction budget. If the owner rep has a good working relationship with the golf course builder, these RFPs will be a simple matter of review; however, when a nongolf course contractor is building a golf course, the RFPs must undergo more detailed review to ensure all line items submitted and their percentage completions are accurate. RFPs also allow for easy review of the sequence of project completions.

Permit Management

In serving as the owner rep, one of the first tasks of the planning/management process is researching the required permits for the project. This is often accomplished by the land planner or developer in conjunction with input from the architect, but it may also require the attention of the golf course superintendent.

The first permits to acquire for the course are the wetland concerns and erosion and sediment (E&S) control plans. The first

Figure 1.9. *Wetland structure construction is possible through permitting.*

requires a wetlands consultant to evaluate the property and identify wetland areas. Today there are many private consultants that serve this purpose; contact the Army Corps of Engineers (COE) for a list for your review. Once wetland delineations have been determined, they can be matched up with golf course or development design plans to see how much, if any, wetland area is impacted by the golf course and if there is proposed fill in this area. The wetlands impact permits can also involve other agencies including the Fish and Wildlife Commission, the Water Quality Division of the Department of Natural Resources, and the Historic Preservation Society. The specifics of the size and quality of wetlands impacted, as well as mitigation concerns, all play a part in the wetlands permitting process. The wetlands consultant can give recommendations for what will be involved in the permitting process before the COE is contacted for official delineation of boundaries and evaluations of permit approval, modification, or decline. The COE has the final word in wetland delineation (Figure 1.9).

An important question to consider here is who will be responsible for overseeing and monitoring the maintenance of the permits? Who will monitor construction activities and ensure they stay within the permit boundaries? Overseeing the wetlands protection, the E&S plan, and the water quality plan sometimes falls to the superintendent as the owner representative. Usually the E&S control plan is installed and managed by the golf course contractor during the construction phases. As this is sometimes pulled out of the contract, however, the superintendent as owner rep must have a clear definition as to what his responsibilities are for E&S control and other appropriate permits in terms of monitoring, measuring, and record keeping.

Wetlands

Your wetlands consultant should be well versed in permit options and support the project interest within the constraints of the law. He or she can also assist the golf course architect with mitigation areas that may be needed on the golf course in keeping with design considerations.

An environmental consultant is used to delineate wetland boundaries and quantify the quality and special considerations of wetlands. The actual boundaries and specifications of the wetlands are then verified by the COE before a permit is issued. The COE estimates that 95% of all permit applications are eventually issued.

The two primary concerns when evaluating wetland permitting are:

1. Is there an "upland alternative"?, and
2. Is there any "net loss" of wetlands?

The actual definition of "wetland" has several aspects that must be evaluated. For example, land having water within the root zone for 5% to 12% of the growing season is considered a wetland environment; hydric soils are another factor, and a dominance of wetland plants still another. This definition includes artificially created wetlands as well.

Potential archeological sites are another important consideration. The environmental consultant should be qualified to give helpful guidance with these considerations and other areas where agencies will potentially interact with a project (30).

Preparing and allocating budget moneys for permit maintenance and monitoring to stay within specifications is critical. Implementation and management of an erosion and sediment (E&S) control plan or a wetlands protection permit is not simple or inexpensive. Daily inspection and maintenance is required to maintain the integrity of the protection system. This routine maintenance and monitoring is more clearly covered in Chapter 7 where E&S control management is discussed in detail.

The last part of the permit scenario involves determining the demands or specifications of the granted permit after the construction and grow-in phase is complete. Sometimes various permits require specific monitoring for a given period of time after the establishment is complete, while other permits require continuous monitoring for an indefinite period. If the superintendent on-site during construction/grow-in does not stay on as maintenance superintendent, he or she is often required to train the new superintendent on permit specifics.

Audubon Program

Participation in the Audubon Cooperative Sanctuary Program should be part of the initial planning phases of the golf course development. Audubon International has programs that involve the golf course from the planning stages of preconstruction to maximize the environmental stewardship in terms of design, grass selections, maintenance practices, and protecting wildlife habitat in and around the golf course. This is an excellent program in which to ensure the environmental stewardship of maintenance operations.

It also assures the new golf course's neighbors that management of water, fertilizers, and control products will be well planned, used with great discretion and as minimally as possible. Wildlife habitats on the course cause very special concerns carefully outlined in the Audubon International program. For example, corridors in and around the golf course are maintained completely natural to allow for "walkways" so wildlife can move to various habitat areas without direct exposure to the openness of the golf course. These types of programs must be incorporated in the initial planning phases and meshed into the overall design.

A golf course applying for Audubon Cooperative Signature Program certification today has the opportunity to achieve the highest certification possible through proper environmental stewardship and sanctuary planning prior to construction. Audubon International provides excellent guidance with design considerations for wildlife habitat. These environmental parameters can be molded into the design concept in conjunction with the architect, so the Audubon Sanctuary Program can be planned from the beginning.

The Audubon Signature Program must address initial course design and development in areas such as:

1. The plan for rough design and maintenance, and how immediate rough areas are blended into existing surrounds
2. Turfgrass species selection and irrigation design
3. The incorporation of IPM programs
4. Environmental impact of water quality and usage and how storm water is collected and managed from the course
5. Natural plant material selection, replanting on the course and plans for future course plantings
6. Wetlands management programs for protection during construction and long-term maintenance
7. The amount and size of internal ponds and the aquatic management plans
8. Planned specialty programs of continued management of the course, such as:

a. the expansion of wildlife feeding areas
b. the expansion of various native birds' habitat for
   their proliferation
c. public awareness programs, specifically for
   schools
d. continued IPM monitoring plans to maintain the
   best management on the course as it matures and
   changes.

The Audubon Cooperative Sanctuary Program is an excellent environmental stewardship program for any new golf course or one currently being renovated. Existing courses can also obtain Audubon Sanctuary status once criteria are met. This leadership program is a natural fit for any new or renovated course to truly expand maintenance programs to their environmental best.

Water Quality/Quantity Evaluation

In the initial phases of golf course planning, the availability and quality of water must be thoroughly researched and determined. Water source evaluations of quantity and quality are done on the basis of the greatest needs of the course which occur during summer stress and grow-in.

Possible irrigation sources include surface water (lakes, streams), groundwater (wells), potable water, or recycled water. To quantify water availability, a water resource study must be prepared which details water quantity, quality, and cost. Until all three are measured, potential irrigation problems are unknown.

Appendix 4 is a peak demand balance sheet bar graph which helps quantify the greatest need versus the availability. If the primary irrigation source cannot meet these greatest demands, then supplemental sources can be evaluated prior to grow-in to prevent shortages. Geological engineers and surveys are necessary if sources or quantities are limited. The turf manager must accurately evalu-

Figure 1.10. *Water requirements during grow-in are significant.*

ate the course's water needs based on the turf species, area of turf to be irrigated, and the normal weekly applications with relation to relative humidity and evapotranspiration rates of that particular environmental climate (Figure 1.10).

Once sufficient quantity has been researched and evaluated, water quality must also be measured. In many parts of the country, water quality is taken for granted because it has always been readily available in supply and in excellent quality. Turf managers in coastal and desert southwest areas, however, view water quality and quantity with a much greater respect.

When sampling water quality, it is essential to duplicate the sample just as it would be on the turfgrass. In other words, if drawing from a large lake and the foot valve is placed 10 to 15 feet below the surface, then take a water sample below the surface, not skimmed off the top. Water quality can vary within the body of water, especially one with little flow exchange. Appendix 5 gives some suggested guidelines for test comparisons of a water analysis and evaluation of the quality of water with respect to various parameters.

Irrigation Water Suitability

When taking water samples, the ideal container is a clean plastic bottle rinsed thoroughly with the water to be sampled before being filled with the sample water. This ensures an uncontaminated sample.

Water suitable for turfgrass irrigation should not contain more than about 850 parts per million salts. Irrigation water above about 2,000 parts per million may injure grass, though some turf species are much more tolerable of soluble salt conditions than others. If salt levels are high, then further testing is necessary.

Irrigation water high in soluble salts is not the only culprit for salt damage. Soil conditions are directly related to the amount of salt that can be leached out of the root system. For example, on sandy soil an irrigation water with an electrical conductivity of 1.5 decisiemens per meter (960 ppm) may provide active turfgrass growth. However, the same water may cause turfgrass injury on a silty clay soil because of poor permeability.

There is a distinct difference between soluble salt levels and sodium levels in irrigation water quality. Roots absorb sodium, moving it to the leaves where it accumulates. Sodium toxicity can take on the look of salt burn on leaves. Closely mown turfgrasses such as annual bluegrass and bentgrass are more susceptible to sodium damage because of the accumulation of the water on a greater percentage of leaf area due to the lower cutting height. Additionally, when exposed to high sodium irrigation water, plant tolerance varies according to species. Hand in hand with this sodium level is another term that is used in water analysis, sodium adsorption ratio (SAR), which should be provided with any laboratory water analysis when water is being applied directly to the soil. Water with an SAR less than 3 is generally considered safe for most turfgrasses.

SAR is the ratio of ion concentration of sodium to the combined ion concentration of calcium and magnesium. In general, an irrigation water with an SAR of greater than 9 can cause severe textural problems in soils with a higher percentage of fine textured silt and

Figure 1.11. Turbidity is water contamination—the same as chemical or salt.

clay separates. In sandy soils, the particle size is larger and a higher SAR can be tolerated without creating permeability problems.

Some irrigation water reports will show an "adjusted" SAR if a high bicarbonate level is present. When higher bicarbonate levels are in irrigation water, it can react with calcium and magnesium in the soil and precipitate out as lime ($CaCo_3$) or magnesium carbonate ($MgCo_3$). This in turn can cause an increase in the SAR. These "red flags" in a water suitability test should be discussed with a water specialist or a qualified turf agronomist to determine the magnitude of the problem and subsequent corrective action—both with water treatment and/or soil amendment (Appendix 5).

Suspended solids in turf irrigation such as clay particles, silt, algae, or even weed seeds can also be a problem. If an irrigation source is "dirty," suspended solids can be pumped onto a turf area in a surprisingly short period of time. One example of an irrigation problem from a river source occurred when a layer of silt and clay approximately one-half inch thick was deposited onto the putting surfaces through the irrigation system in a 15-month period (Figure 1.11).

The problem must first be identified, but a suspended solid problem can be corrected with filtration at the pump station. A Reverse Osmosis (RO) water treatment is effective both in cost efficiency and water cleanup. This has been especially beneficial when brackish water is available rather than ocean water, but both are candidates for this process. RO facilities can effectively filter high concentrations of salt from water and produce a suitable quality for irrigation. Coastal or island courses often lack another alternative with adequate quantity. A variation of this process, known as membrane softening, has also made water with high hardness or alkalinity available. Some courses have saved on water costs as compared to buying potable water for irrigation.

Evaluating water quantity, quality, and availability for a new golf course necessitates initial testing. Golf course superintendents have often been severely restricted by water availability when at first it "seemed" water would not be a problem. Water quality details are more thoroughly discussed in other references such as *Wastewater Reuse for Golf Course Irrigation* (23).

CHAPTER TWO

THE QC/MANAGEMENT ROLE AS OWNER REPRESENTATIVE

Owner Rep Input with the Architect/Builder

Hopefully the golf course superintendent will be hired early in the project's planning phase of construction. The golf course superintendent can control the quality of finish construction and serve as a qualified pair of eyes as the owner representative, and is a valuable asset in the planning process by helping to make decisions and setting specifications for the golf course (Figure 2.1).

These specifications include: grass selection, greens mix, greens construction type, tee construction type, drainage, erosion and sediment (E&S) control monitoring setup, and material approval. The superintendent should also be active in determining the sod budget, providing suggestions for cart path and bridge width and material selection, and of course providing input on the irrigation system's design details and capabilities. The superintendent helps owners understand the architect's schedule for visiting,

Figure 2.1. *Poor construction technique is costly and reduces playing quality.*

signing off on design features and shaping as the golf course construction progresses, and can complement this coordination of architect, builder, and owner.

There are many advantages of using a qualified golf course builder. Selecting a nongolf course builder significantly increases the quality control (QC) duties of the golf course superintendent as the owner rep. When a qualified golf course builder is utilized, the superintendent's main duty is to monitor the above outlined areas and help coordinate specific site needs, including the permit monitoring process, E&S control monitoring, and providing input to the golf course architect for grass selection and maintenance of design. Material selection, approval, and quality monitoring are also key tasks.

With the magnitude of golf course construction activity today, many golf courses are being built or renovated by nongolf course contractors. Many grading, landscape, or earthmoving contractors have obtained golf course construction or renovation jobs with very little, if any, golf course construction experience. Such a contractor may move soil effectively and efficiently, but the intricacies of golf course construction require a trained and qualified golf course contractor to carry out all the details required for maintainability and playability such as proper preparation of seedbeds and proper shaping of slopes to accommodate mowing traffic and play. Oftentimes with a nongolf course contractor, grades are more severe than desired and grades or the shaping on slopes is less contoured to tie into existing terrain than when done by a professional golf course shaper.

The Golf Course Builders Association of America (GCBAA) is an excellent reference source for locating qualified golf course contractors and in helping with contractor recommendations in your locale.

Two of the greatest QC activities that the golf course superintendent has during construction are monitoring the quality of irrigation installation, and monitoring greens construction methods and materials. As the construction process draws to a close, the golf course superintendent gives attention to finish grading techniques in seedbed preparation, preplant material selection, grassing, and beginning the grow-in.

Irrigation Design

This book will not detail irrigation design and engineering of the system, but it is important to discuss some system "features" for maximum efficiency. The first design consideration is the irrigation system around the putting green complexes.

Irrigation needs of high sand profile greens and adjacent topsoil slopes differ significantly. This can be a maintenance nightmare which is magnified during grow-in. The green complex slopes often become so drought stressed during the summer as to prevent overwatering on the putting surfaces. If additional water from the putting green is put on the perimeters to eliminate drought stress, then overwatering and/or algae problems can result on the putting surfaces. If slopes are sodded before greens are seeded, then mowing slopes can become nearly impossible from the green's grow-in irrigation. This is magnified if a dual system is not in place around the greens. Hand syringing is an essential part of putting green management, but it is often limited as to how much of this need differential can be made up.

A split system around the putting greens more efficiently supplies the water needs of the perimeter slopes and the high-sand greens and collars. Although the split system or dual application system around putting greens adds some cost to the system, the long-term benefits in turf management and better playability far outweigh the initial capital investment. Appendix 6 illustrates a typical split system and the flexibility it offers.

Systems may be part circle or one full circle and the other part circle. Full circle heads are best to water the collar and putting green, while the partial circle system supplements the application of water to the green slopes. This allows different water regimes to be applied, which is especially critical during establishment, drought stress times, and in the southern overseeding market. This type of system is simpler for installation and maintenance, and provides the maximum flexibility between the two systems versus having both systems partial circle. The difference in application rates between heads is also more consistent, since all heads have the same rotation. There are few instances where this system would not be the most efficient. Almost without fail, the green perimeter areas need more water overall than do the putting surfaces although the putting surface is a high-sand profile. Properly built high-sand profiles have excellent water retention and require deep and infrequent irrigation cycles supplemented with hand syringing of dry spots.

Location of the quick coupler around each green is another irrigation design consideration. Figure 2.2 illustrates the most common installation of the quick coupler—one adjacent to an irrigation head. The quick coupler is best located on the back of the green so the hose can be easily moved and kept out of play when syringing and it should always be put in the same location; for example, always at the back left or back right of each green.

For future irrigation system management of the golf course it

Figure 2.2. Irrigation system installation with quick coupler.

is wise to hire an irrigation technician and allow him to be a crew member for the irrigation installer so he will learn the system installation as thoroughly as possible. Many irrigation contractors utilize this program with golf course superintendents and are very flexible about hiring the technician as a crew member during the installation process. This person is also an extra set of quality control eyes for the golf course superintendent (Figure 2.3).

The owner rep should be a help and not a hindrance. Qualified golf course builders do not need golf course superintendents to help them with construction techniques, drainage engineering or installation, or scheduling of job duties and managing E&S programs. However, they do welcome assistance with material selections, grassing selections, and other areas related to agronomics. Simple things during the construction process, like helping to locate bury holes for debris and topsoil placement locations are quality control jobs for which the superintendent can provide welcomed input to the golf course contractor. Then all are providing expertise for an improved finished product instead of a confrontational situation where everybody is trying to be "boss."

Figure 2.3. Quality irrigation installation is an absolute necessity.

Turf industry professionals do not have extensive construction backgrounds, just as construction contractors do not usually have extensive turf backgrounds. As an owner representative, allow everyone to do their job in their area of expertise for the best finished product.

The superintendent can provide excellent specification input for the environmental aspects of golf course development. The most common of these are grass selection, greens construction type, and irrigation design extensiveness. However, many other areas require well-planned specifications for golf course construction, depending on the planned use of the golf facility.

In the planning process, the superintendent can coordinate the level of quality of the finished product in relation to the budget. It is common for owners to build a golf course with high maintenance quality standards in mind, only to later discover that the required maintenance budget is much higher than they had imagined. A golf course superintendent can provide an estimated budget for the level desired, both in terms of maintenance quality and the design of the golf course. A golf design with extreme features and undulations requires much more hand mowing than does one with softer features. Consequently, the maintenance budget must reflect these design features. If the budget is much higher than the owners anticipated, then some of the design severity can be softened before shaping begins.

The amount of play anticipated for a golf course should also affect the design specifications in green size, tee size, width of fairways, the severity of outer roughs, and the amount and severity of bunkers. A daily fee course designed for high traffic and the higher handicap golfer would have softer features and would need larger tees and greens than a private club designed for average play. The term "large greens" is vague because there are many large putting surfaces designed and built today with very few cupping areas because of severe undulations. Size and undulations must be considered when designing a high-play golf course to have maximum cupping area over the green surface. Putting greens should average 5,500 to 6,500 sq. ft. minimum with a reasonable amount of cupping area over the putting surface. However, if 42,000 to 45,000 rounds or more per year of golf are anticipated, then greens should be 6,500 to 7,000 sq. ft. minimum to provide better traffic distribution through more cup locations over the entire putting surface.

Tee size is a critical design factor when estimating the projected rounds of play. Tee size should be 100 sq. ft. per 1,000 rounds of golf a year and 200 sq. ft. per 1,000 rounds on par 3 tees or tees where an iron shot is played. Additionally, those tee sizes must be proportioned equally to the amounts of play by the respective golfers. For example, the regular men's tee will receive the most traffic and proportionately must have the same amount of teeing area for that golf hole. Careful input by the architect and superintendent can ensure that the proper teeing size proportions are designed into the different tee box locations on the golf course (Figure 2.4).

Tee width, not just length, is a design factor that should be evaluated for golf course development. Tees are often long and narrow, which does not allow lateral movement of the tee markers. A long, narrow tee eventually shows wear in the middle and to the right because of the predominance of right-hand golfers. Conse-

Figure 2.4. Small tee surfaces create poor playability.

quently it receives very little use on the left side, compared to a wider tee which allows lateral traffic flow.

Sod Budget

All golf course construction/renovation projects have a line item for sod. Golf courses under construction across the country average a sod budget of about 400,000 to 500,000 sq. ft. Many developers see this as excessive and a good place to cut the construction budget, but when "feature" areas of the golf course are measured, 500,000 sq. ft. of sod is not an extravagant amount for almost any project.

The optimum goal for sod utilization is to cover all green complex, tee and bunker slopes. Additionally, a small area around catch basins or drainage swells are excellent choices for sod. Tee tops and putting surfaces are not the best choices for sod unless the budget allows because these areas are easily established from seed

Figure 2.5. A carefully planned sod budget is invaluable.

or sprigs. We will discuss this more when addressing the use of soilless or washed sod.

Within the sod budget the grow-in manager must carefully measure and mark particular erosion problems during construction. These key areas for sod prevent future damage. Even if it means increasing the sod budget initially, sodding washed areas will save a significant amount of money over the course of grow-in and maturity. Once a washout is repaired one time, sodding is the most economical way to establish a wash area, compared to repairing washes—adding soil, raking smooth, compacting, cleanup of the eroded soil, reseeding or sprigging, and mulching. This is one area where sod becomes an economy instead of a luxury. Figure 2.5 is a perfect example of a properly managed sod budget on a golf hole to maintain feature shaping and minimize erosion without excessive sodding.

Total course sodding today is gaining popularity from an environmental view and a more rapid opening. Solid sodding of greens, tees, fairways, and immediate roughs is very costly but results are instant and play can begin much quicker. The cost of sod versus

Figure 2.6. *Total sodding is a grow-in option.*

revenue must be carefully weighed. Generally speaking however, tee tops and putting surfaces can be established with smooth, firm surfaces just as quickly with seed or sprigs as with sod, at much less cost. These are the least needed sod areas because of the sand base and level surfaces, unless total sodding is the preference (Figure 2.6).

Cart Paths

Details to consider when evaluating cart paths for a golf course are the materials to be used, the width, providing turnouts or bypass widths at tees and greens, the use of curbing, and the proper routing of cart paths. The architect will have the greatest influence on cart path routing, and is most qualified to provide proper routing to maximize the aesthetics of the cart path in blending into surrounding landscape. He or she will also provide the best means of routing the cart paths for playability of the course and the projected traffic flow.

The management team should decide on the material choice for the cart paths. Concrete with a fiber mesh additive is the most commonly selected material today for cart paths but asphalt paths still provide excellent service as a cart path material in many locations. Some aspects such as cost and color of the materials should be weighed to consider the owner's preference.

The recommended width of 8 ft. for golf course cart paths will accommodate all maintenance equipment and carts. Seven-foot paths have been installed to save some money over 8 ft., but 5 or 6 ft. wide cart paths should always be discouraged as they suffer significant wear along the cart path edge due to traffic running off the edges.

Curbing along cart paths is a critical part of the path system, especially at tees and greens to inhibit badly worn traffic patterns. Golfers pull off the edge of the path with the carts just as we do with cars along the edge of the road. Curbing helps prevent these damaged areas and has also been used effectively to channel storm water into basins. Considerations such as curb height and construction relating to path/curb tie-in should be questioned by someone with curb construction experience. Qualified golf course builders will utilize experienced cart path subcontractors to detail aspects like this as well as the proper subbase of the cart paths (Figure 2.7).

One last cart path planning detail is a crossing sleeve installed under paths in low areas. A 4–6″ PVC pipe is best, usually placed 12″ below the concrete/asphalt. This allows future drainage or irrigation crossings without cutting the paths. Locations are recorded on field notes and later transferred to asbuilts. This step is inexpensive and can save significant costs in the future.

Finally, when evaluating traffic on the course, planning for easy access by service trucks to the pump house cannot be overlooked. Considerations include:

1. Pump station service trucks
2. Power company accessibility to the transformer

Figure 2.7. *Cart path quality requires a competent contractor.*

    3. Service to the pump station by the fertigation
       supplier if fertigation is utilized.

    Lack of service roads could cause significant damage to the
golf course. Reviewing this access need in the planning stages,
however, should eliminate this problem.

    Like cart paths, retainer walls and bridges are important
construction amenities that require proper design and material
planning. Figure 2-8 is an excellent example of proper construction technique and materials, engineered for stability and load capacity. Marine grade treated lumber of adequate sizing was used
and pilings were driven to "refusal" in the ground for a proper
foundation.

    "Deadmen" is an anchoring term referring to the tiebacks for
vertical stability on retainer walls. Stainless steel cables anchored
with concrete footers or treated timbers are used very effectively.
Adequate footer pilings and deadmen anchoring are the key for retainer wall security.

Figure 2.8. *Proper retainer wall construction.*

Managing the Operation During Grow-in

Budgets

The grow-in budget receives little attention during the entire planning process. Owners and developers must be made aware of the grow-in and first year maintenance expenses early in the construction stages so money can be allocated.

Often, the grow-in budget is planned for at the end of construction, coinciding with a time when cash flow is at its worst. Furthermore, many owners believe, incorrectly, that the grow-in is part of the construction budget. Appendix 7 is a construction budget overview done in 1996 to identify average cost percentages for construction in various categories. Note that grow-in is not identified as a part of construction.

The grow-in budget and the first year maintenance budget have very specific categories and needs unlike normal maintenance budgets. Both have job duties which must be well planned for in both materials and labor. Developers or owners may consider holding off on many projects early in the phase to spread the cash flow, but certain jobs must be carried out in the proper sequence within the grow-in and first year's maintenance.

Appendix 8 details several specific budgetary areas requiring early planning in course development. These include examples of:

1. the grow-in budget
2. the grow-in equipment budget
3. the maintenance equipment budget
4. the mechanic shop inventory
5. the initial course stocking
6. the hand tool and accessory budget.

These suggested budget categories indicate the necessity of planning for needed expenditures early in the development phase. These categories indicate the detail required, but each must be altered to individual needs.

In this book, normal maintenance budget operations or equipment inventories are not discussed other than to provide *planning* suggestions. Design, grass selection, or maintenance level can vary the maintenance budget significantly on any given golf course. Involving the superintendent early in the construction process will help bring together expectations of the architect and the owner/developer, and help the owner understand the maintenance and expenditure needs of the golf course. Design and maintenance must concur with the owner's intentions <u>before</u> grass is planted.

For example, bentgrass fairways provide a higher quality playing surface than bluegrass or ryegrass but are also more costly to maintain. Is the owner aware of these additional costs? Is the increased playing quality of the bentgrass fairways worth the extra cost to the owner? Is the owner aware of the cost of extra hand mowing of the steep slopes in the green complexes and bunker faces and willing to pay for this labor cost, a continual expense throughout the life of the golf course? What about the differences in cultural programs and control product needs?

These are just a few examples of questions to be weighed by the owner and why golf course superintendent input at this phase is so important. After considering plans, specifications, details, and desires, the superintendent can develop and accurately estimate a projected maintenance and equipment cost for the golf course.

Though the golf course maintenance costs are considered, what about the time and budget from planting until opening? The finish construction and grow-in phases are a poor time for budgetary surprises. The budgetary considerations and ideas in Appendix 8 must be carefully weighed and planned for during construction to eliminate the confusion that so often happens at grow-in.

Efficient budget preplanning includes using used equipment for grow-in which saves money without sacrificing quality. Many distributors have a continuous inventory of used equipment from trade-ins that allows a golf course to acquire the majority of needed equipment for grow-in from the used market. Early planning by the superintendent is necessary to reserve used equipment for use when grassing begins.

Many superintendents purchase new green and tee mowers and then buy used reels for the machines during grow-in. This protects a new piece of equipment from harsh grow-in conditions; then afterward these used reels are rebuilt and utilized after topdressing to protect the normal maintenance reels. Either scenario is smart management to achieve the greatest use from today's demands of golf course equipment.

The one area not listed in Appendix 8 is the cost of the maintenance facility. As outlined, planning for the maintenance facility early and beginning construction to target occupancy by grow-in can not be overemphasized. The grow-in is a critical time for proper equipment maintenance and storage, and often maintenance facility construction has not even started when grow-in begins. Interior details of the maintenance facility do not have to be complete by grow-in, but the exterior should be finished so the building is ready for storage and maintenance.

Maintenance facility design and construction should be coor-

Figure 2.9. *A well-planned wash station.*

dinated by the golf course superintendent through the owner to ensure facility readiness by the time grow-in begins. Maintenance facility details are not discussed in this book other than to reiterate the absolute necessity to plan early for construction. Many courses are forced to work from temporary facilities well into grow-in when the maintenance facility should have been completed at the <u>beginning</u> of establishment.

The wash/loading zone of the maintenance facility requires special attention to environmental standards. Figure 2.9 is an example of a wash station constructed at Pinehurst Golf and Country Club, Pinehurst, NC. Clippings are channeled into a catchment and then removed as often as the mowing operation requires.

The rinsate water is cleaned through a charcoal and sand filter and stored in an adjacent tank. The water is then used as irrigation on an area of the driving range for disposal. As Brad Kocher, certified golf course superintendent said, "Why not utilize one of the best filters—turf". Similar systems can be adapted to meet wash water cleanup concerns. The local state university or an environmental consultant should be dependable resources for permits and requirements.

More important areas of concern for maintenance facility planning, other than local codes and permits, are outlined in Appendix 9. Make every effort to expedite maintenance facility planning with the owners/developers. Managing grow-in from a tent is not an efficient situation.

During the initial construction phases, the superintendent can identify the best maintenance facility location, do site planning, permitting, and begin the construction of the building in the proper sequence. The superintendent can also research the size and needs of the maintenance facility to match the golf course operations. An additional consideration to weigh when determining maintenance facility size is whether or not the landscape operations for the project grounds are to be operated out of this maintenance facility. Some specific details in size and design should be incorporated into the maintenance facility when two different operations share a common building. Early planning for these considerations will greatly improve the efficiency of operations and over the long run will save significant money in facility cost versus retrofitting to increase the size of a building that is too small.

Managing the grow-in operation requires a strategic understanding of all phases and how job duties, concerns, "red flags," and needs change as the grow-in progresses. He must be prepared to adjust schedules to match changes in weather, planting windows, equipment problems, and of course, the unexpected.

Key management areas to define, understand, and then plan/budget for include:

- Proper evaluation of labor needs
- Job duties
- Delegation of those duties
- Scouting—IPM and grow-in progress
- Irrigation management
- Fertilization management

- Pest control
- Mowing management
- E & S management
- Equipment care under stress circumstances
- Anticipating the unexpected

The overall grow-in budget is usually set up on a six-month basis, although this time frame can vary depending on the planting window of the golf course. For example, if three months of winter fall within the grow-in period, which extends the actual calendar time to nine months, the grow-in itself should still be calculated on a six-month basis, with the additional three months of winter budgeted for labor and special projects. As will be discussed later however, this winter time period occurring during grow-in has created some problems for superintendents because of the time duration versus the *actual growing season*.

The grow-in labor force for an average 18-hole golf course in the transition zone usually consists of 11 to 14 employees, <u>not</u> counting the superintendent, the assistant superintendent, and the mechanic. This number varies significantly if areas such as landscaping are included in the grow-in budget and duties of the golf course staff. The severity of design influences the amount of hand labor involved, predominantly in mowing and trimming, and this too can vary the number of grow-in employees (Figure 2.10). Understaffing during grow-in is one of the worst mistakes made by many golf course developments because a rapid and efficient grow-in saves money many times over skimping on the grow-in budget. An adequate work force and equipment cannot be emphasized enough during grow-in, from planting until opening.

Appendix 10 is an example schedule of a basic day as a grow-in manager. The job duties are numerous and priorities change weekly. If a superintendent does not have some idea of how these will be accomplished, their course of progression, and what is required to meet the goals, he will be lost from day one. Waiting until the grass is planted is too late to organize. *Planning* and *planting* cannot happen at the same time!

40

Figure 2.10. *Approaches are a place where hand mowing is commonly needed during grow-in.*

Flexibility and effective in-field management are the keys to successful grow-in management. Scouting the efficiency of all operations and continually making adjustments according to results produces an aggressive versus a mediocre grow-in. Flexibility is key because constant changes are a major part of the grow-in process. The unexpected rainstorm, the fall armyworms outbreak, an irrigation pump that goes down, the fertilizer shipment that is delayed, or an equipment breakdown are only a few of the problems that can arise. Are you prepared for these and have a Plan B to compensate? Planning ahead is more than just helpful. Understanding the process thoroughly allows for this flexibility (Figure 2.11).

An assistant superintendent is invaluable during the golf course grow-in. He should play a major part of the in-field monitoring but usually has little experience with finish construction and grow-in. Careful training is needed to give an assistant a "crystal ball" observation of what the grow-in will entail so he is better prepared for what to expect out of <u>planning</u> rather than <u>experience.</u>

41

Figure 2.11. Potential thunderstorm damage must have a preplanned cleanup
program.

The experience and maturity come with greater productivity if an
assistant has an idea of what to anticipate. It is easy to see how
grow-in training can compensate quite a bit for experience.

The same is true with the crew. The mower operators, for example, must understand the damage an out-of-adjustment or stuck
reel will do and stop mowing immediately. Will the mechanic make
periodic checks in the field to see how operations are going? This is
another branch of the in-field management strategy.

Daily monitoring of primary agronomic programs is the
largest part of management responsibilities. Evaluating watering
and fertility efficiency and proper readjustments according to performance are of premium importance. The specifics of these agronomic programs are detailed in Chapters 5 and 6. Monitoring turf
health and growth regularly allows effective implementation of
agronomic programs.

IPM scouting balances the daily monitoring duties of the
management team. Aggressive evaluation of pest activity keeps pesticide applications minimal by spot treating only areas of activity.

Threshold levels are more efficiently known by regular and consistent scouting and not periodic scouting.

As outlined in Chapter 1, continued weekly meetings with owners/developers and management team will keep everyone informed on the critical path and progression of the golf course. This drastically reduces "fire fighting" in management meetings.

Punch Lists

Any construction project has a list of items that need repair, replacement, or to be addressed as it nears completion—a punch list. The construction of a golf course is no different. The owner rep plays a key role in making sure this punch list is complete and carried out satisfactorily. The superintendent should develop his/her own punch list which should be discussed with the owner, the architect, and the builder. Once everyone agrees with punch list items, it is easier to accomplish and prevents disagreements from occurring at the project's completion.

A punch list is most effectively and fairly developed by a walk-through of the golf course with the owner, owner rep, architect, and the golf course builder. Notes are taken and the scope of work is agreed upon during this walk-through. After the walk-through is completed, the punch list is typed up and redistributed to all individuals so, again, everyone is in consensus with the definition and length of the punch list. As an owner representative, a completion date of this punch list should be required to expedite the process and allow for better grow-in scheduling. This process is the fairest and most productive to all involved.

Laser measuring the golf course is an important part of the development stage which usually occurs at the latter part of grow-in. As outlined in the critical path in Appendix 1, measuring the golf course must be timed to allow the results to be returned, distance markers and hole signage to be ordered, received, and then installed on the golf course. Many developments have had to delay opening dates because the golf course amenities were not yet in place.

Hole signs and distance markers should be put in about one month before opening, which gives a reasonable time to complete their installation and allow for any adjustments or repairs to be done. In the initial planning phases, golf course amenities should be studied because if amenities or hole signs are chosen that require a particularly long period of time for construction, then this time frame must be considered in the development process.

CHAPTER *THREE*

AGRONOMIC CONCERNS OF
FINISH CONSTRUCTION DETAILS

Many construction details must be properly handled and well monitored prior to grow-in which will drastically affect the quality of turf on a short- and long-term basis as well as greatly affect playability. Settling of trenches in fairways and around greens is a great example. Proper compaction in irrigation trenches is an absolute must, an example which is displayed in Figure 3.1. Construction during drought periods can reduce the ability to compact trenches because dry soil will not compact effectively. In such cases the superintendent would be wise to wet the trench soil by hose from a water wagon or other means after trench backfilling. Putting moisture in the soil before compacting will eliminate future settling, which is especially important for trenches within playing areas. Let's review several of these QC issues for finish construction.

Figure 3.1. *Trench compaction is an important detail of quality construction.*

Greens

Greens construction methodology has been a major issue of debate for years, but in the last few years many variations claim to be superior in method and performance. The USGA Green Section spec has been the standard for high-sand green construction over the last 35 years. Refer to the March/April, 1993 *Green Section Record* (22) issue regarding updates in specifications for construction methods and materials.

Straight-sand green construction and various types of USGA "modifications" have come onto the market with many claims and expectations. When considering variations in high-sand green construction, the first demand should be to see the university research that substantiates this type of construction. Most non-USGA recommendations on high-sand construction lack university backing, but instead have someone's opinion on their short- and long-term performance. This is not sufficient documentation.

In this chapter, we will refer simply to the concerns of high-sand greens. The details and concerns outlined will pertain to most any type of high-sand construction, except when differences are noted. We will discuss the primary mistakes that have been made

Figure 3.2. *Representative sampling is a must.*

in the field over the years in greens construction and key QC issues that must be monitored closely to ensure the integrity of the desired and designed putting green construction.

The methods and materials used in putting green construction probably require as much or more QC monitoring than any other single area besides, perhaps, the irrigation system. Careful checking and monitoring of this construction phase will yield superior putting green performance for an indefinite period of time. When high-sand greens are properly built, correct materials selected, and then the greens properly managed, green integrity will last indefinitely.

The base ingredient of green construction is sand. Many golf courses are built today using a local sand which is oftentimes unacceptable or the sample taken was not representative of the source. Figure 3.2 shows the proper technique of sampling a large sand pile to achieve a representative sample. This is best done with a 2-inch PVC pipe that is pushed into the sand pile (usually 2 to 3 ft.) at about chest height and then removed with a representative core sample. This is emptied into a large bucket and this procedure is repeated until the entire perimeter of the stockpile is sampled. This sample is then thoroughly mixed and one gallon sent off to an approved soil testing lab for analysis. The superintendent should keep another gallon of the sample for future reference. The soil amendment of choice should also be sent to determine its performance with the selected sand. Likewise, gravel choices are sent to evaluate their compatibility with the sand, and thus, whether the intermediate sand layer is required.

When discussing the physical characteristics of a root zone mix, be it straight sand or with a physical amendment, the turf manager should ask the laboratory questions about the pore space distribution, as this will allow him the greatest understanding of agronomic characteristics of that root zone. The grow-in manager who understands the agronomic performance of soils on the golf course before planting can plan for efficient irrigation and fertility programs once the germination and establishment begins.

Infiltration rate probably receives the greatest attention when evaluating a physical soil test report. However, the macro and micro pore space percentages are equally, if not even more important numbers to understand because they reflect the performance of the root zone mix more versus infiltration rate per hour. Macro and micro pore space balance dictate a sand's air:water ratio balance—how wet or drought-prone the profile will be. Making sure this balance is correct produces a field capacity, capillary water availability, and a soil moisture wilting point most conducive for turfgrass health. When a root zone is evaluated with a USGA approved physical testing lab, the superintendent can discuss these criteria with lab personnel to understand a root zone's performance characteristics before it is selected. This is an absolute must if different sands are being compared based on cost and/or availability. Local sand sources are numerous in some locales and accurate sampling, evaluation of physical makeup and volume supply can prevent a nightmare or save significant money in materials.

For example, if a root zone mix meets specifications and is at the upper end of the micro pore space scale, the lower end of the macro pore space, and has a higher percentage of fines, then this mix will have greater drought resistance. Furthermore, grow-in irrigation management with this mix will probably require one or two less irrigation cycles per day during germination than would a root zone mix high in the macro pore space range and lower in the micro pore space range.

As outlined earlier, material selection must be carefully researched on any high-sand profile to avoid excess infiltration rates, coarseness, and provide a proper balance of macro or micro pore spaces. California type greens, like USGA specs, have "modifications" which have no substantial research support in most cases. Avoid these "modified greens." Explain to owners the theory behind properly constructed high-sand greens. Their long-term benefits can only be fully experienced with proven material selection and construction technique.

Straight-sand greens construction should be mentioned because of its significant popularity. The California method has been widely used because it is less expensive to build. Some climates have more difficulty with straight sand construction than do others because of water retention. In the first two years a high-sand green experiences greater infiltration rates and a tendency for leaching and localized dry spots. A root zone mix without an amendment magnifies this tendency.

The management characteristics of any style high-sand green construction chosen is largely affected by the quality of the mix selected. This not only includes the choice of sand and its physical characteristics, but also the qualities of the selected organic matter or other physical soil amendment and its characteristics with that sand. Research shows that organic matter performs differently with different sand, and ceramics are no exception. The automatic incorporation of any soil amendment to sand does not necessarily mean improved physical performance of that sand. Not only the behavior of that amendment, but the correct amount proportionally must be evaluated with the chosen sand.

On-site mixing or blending soil amendments with sand using a rototiller after the sand is put in the green cavity is a practice still widely used today. Unfortunately, it is nearly impossible to uniformly blend an organic matter or soil amendment with sand in such an operation. Figure 3.3, Plate 1 is an example of a mix blended with six replications using a high RPM, tractor-mounted tiller. All six replications were in a different direction. Note the "layer" at the downward reach of the rototiller, and below that is the straight sand. Also notice the peat moss clumps that exist because of the lack of shredding. This demonstrates the problem with on-site mixing, and though elaborated in other publications, it is worth mentioning here to reinforce the fact that on-site mixing is not an acceptable substitute. The small savings incurred by on-site blending in no way justifies the cost of the golf course construction or renovation and the long-term effects of performance. As with proper sand selection, blending techniques greatly affect the physical behavior of any type of high-sand construction mix.

Greens mix depth is another component critical to the performance of the profile and the uniformity of the putting green. Variable depths of greens mix within the profile will destroy the uniformity of water movement and soil moisture in various areas of the green, thus creating an environment that requires significant hand watering. The technology available today with irrigation system design and our knowledge of proper greens construction allows almost total control of green construction performance. To build a green with a varied root zone mix depth which creates moisture uniformity problems, reflects either very poor quality control or last minute surface contouring changes by someone who does not know their job!

The perched water table effect is based on a <u>uniform</u> 12-inch compacted root zone mixture depth, so the movement of water and the percent moisture of the mix is kept consistent on low areas or high areas of the green. This is why the subgrade (although less critical), gravel layer, and/or coarse sand layer should be at the <u>same</u> contouring as the finished grade of the green. Owner representatives should add a paragraph in the contract to have the architect sign off on green cavities once the subgrade or gravel blanket is completed. This helps ensure that putting greens are then built with uniform layer depths, so the agronomic and physical performance of the putting green root zone is maximized. <u>Slight</u> modifications of surface contouring can be made with the underlying gravel blanket, as this would not affect the perched water table characteristics if the gravel layer, for example, ranged from 4 to 8 inches

Figure 3.4. *The gravel base should conform to finish grade.*

deep to accommodate some last minute surface contouring needs (Figure 3.4).

However, varying the greens mix depth from 9 to 20 inches deep, which can happen with last minute design changes, absolutely destroys the beneficial and uniform function that the perched water table concept produces. Greens mix *must* be uniform over the entire putting surface—no exceptions.

One mistake often made at this green construction point is with the placement of the gravel into the cavity. Figure 3.5 is an example where a truck crushed the drainpipe when allowed to drive onto the drain field even though the truck went "across" the drainline. The gravel blanket will not carry the weight to protect the pipe.

Another issue that often arises in evaluating greens construction is determining need for the vertical plastic wicking barrier around the green cavity. Figure 3.6 is an example of this installation during construction. Though often promoted as a prevention for tree root migration into the cavity of the green, this is not the primary function of the barrier. When a high-sand mixture is put di-

Figure 3.5. *The result of heavy traffic directly on the gravel blanket.*

Figure 3.6. *The vapor barrier is important agronomically.*

rectly in contact with a high organic or heavy topsoil such as a clay or silty clay loam, two soils adjoin with completely different affinities for water. The topsoil will wick water from the high-sand root zone, creating a very dry perimeter barrier around the outside edge of the greens mix core. Desiccation is a constant battle.

Vertical wall, cavity construction is usually a preferred method by contractors because it is easier to build, and is superior because the cavity of the green is more clearly identified. Although the topsoil and greens mix can be tapered and feathered together to eliminate this vertical barrier need, the vertical barrier is a preferred method because of the exact identification of the cavity and its subsequent performance for turfgrass growth. The cavity will usually accommodate the putting surface and the collar.

It is critical to identify the core boundary when the greens mix is installed with surveying flags because it determines the performance of the putting green collar when installing the green perimeter sod or the collar sod, regardless of the construction method. Exact identification of the green's core by flagging ensures that the collar is either completely on the greens mix or completely off the greens mix in the adjacent topsoil, whichever is chosen at any particular site. The worst management scenario occurs when the core boundary is not identified and the collar can actually be established partially on topsoil and partially on greens mix. This creates water uniformity and wilt problem nightmares in the collar, and during summer stress periods the collar can be more difficult to maintain than the putting surface. Core identification with flagging and contouring with sod installation is an important QC job for the owner rep (Figure 3.7).

Careful floating and firming of the greens mix is the last attention to detail needed prior to planting. Figure 3.8 illustrates an attempt to apply preplant soil amendment materials on a greens mix that is not firm enough to accept such traffic. If this is the case, apply preplants and then do final firming and floating. The owner representative must make sure, however, that the preplants are not incorporated too deeply into the soil but remain at the surface to benefit the newly establishing turf.

Figure 3.9 shows a properly floated green prior to seeding;

Figure 3.7. *Flagging the green core ensures accurate identification.*

Figure 3.8. *Firm greens mix at planting prevents headaches later.*

Figure 3.9. Proper root zone floating/firming.

notice the surface firmness. As discussed earlier, firming the soil mix with irrigation and properly floating prior to seeding is critical. The firmer the surface, the more intact surface contours will be during grow-in and establishment. Furthermore, much less topdressing, rolling, and aerifying will be required by the superintendent to smooth surfaces. When putting greens are firmed prior to seeding, only rolling and topdressing are necessary to provide great playing conditions after proper grow-in.

When applying preplant fertilizers to a new high-sand root zone mix, evaluate how much tracking occurs from the spreader. About .375″ should be the limit for tire rut depth. More than this requires additional firming before final floating. If it is necessary to add preplants or soil amendments to a mix that is softer than desired, do not completely fill the hopper. Sometimes only 30 to 35 lb. should be put in the hopper at a time because the extra weight increases the problem of tracking and rutting.

The greens mix staging and stockpiling area may seem like a small issue, but it is a critical management area for an owner representative. The contractor needs an area easily accessible by trucks, large enough to accommodate the blending operation (if it is to be done at the site) and accessibility from the staging area out to the green sites. Ideally, the staging area should be completely firm and compacted, preferably with a solid base and a slight slope pitched away from traffic flow. This removes runoff water that accumulates in the huge stockpile of mix. If the stockpile area slopes out into the major traffic area, it creates a constant mud problem because of the tremendous amount of water draining from the mix pile.

Contamination of the greens mix pile from underlying soil is a danger during mixing and distribution. Eight to twelve inches of mix should be left as a base on the floor and the loader's bucket should never go below this so it can serve as a buffer to protect the stockpile from contamination. This prevents the bucket from getting into the underlying soil and/or gravel and redistributing it through the pile as it is reconsolidated to improve loader handling.

Many owners may say they cannot afford to waste this much mix because of the expense. Theoretically, however, only a small amount of mix is wasted when the operation is complete, but the contamination of an entire pile as shown in Figure 3.10 puts this cost into perspective. Furthermore, this mix base is not wasted as it is well suited for topdressing birdbaths and trench fill areas in fairways, immediate roughs, and approaches.

The greens mix will never be purchased any less expensively on a per yard basis than during the mixing operation. Therefore, grow-in superintendents are wise to budget for additional mix to be stockpiled after the greens mix is put in place, to serve as topdressing for the greens for the next one to two years. This ensures absolute consistency with the construction mix and will be the least expensive way of obtaining the initial supply of topdressing.

A final consideration during green construction is a design accommodation for forced air introduction/evacuation through the drainage system. High volume pumps are successfully used to remove additional water from the profile or force air back into the root zone, creating aeration at the bottom.

Portable units are utilized efficiently when a "quick coupler" outlet is installed in the green drainline outfall. Moreover, the drainline system also has some additional loop tie-ins to allow for

Figure 3.10. *Stockpile contamination must be prevented.*

better distribution from the pump. The manufacturer can provide the spec accommodations for this system at construction and it is a very modest cost increase except for the small amount of additional materials. This drainage system modification should definitely be considered.

Tee Construction

Tee construction methods have been the source of much discussion in the last few years of golf course development. The trend has been to cap tees with 4 to 6 inches of compacted sand/sand mix to create a better, more compaction-resistant, growing and playing environment. This is an excellent scenario for agronomic and playability improvements, but in most cases, the wrong sand is chosen for this use. What is the correct sand?

Using a sand similar to that approved for green construction because it is classified as "an approved golf course sand" is often

Figure 3.11. *Sand capped tees can have significant agronomic advantages.*

thought best. Though this may be true for USGA spec greens construction, there is a completely different need when constructing a 4 to 6 inch compacted cap on tee tops. The sand best suited for tee tops is a fine sand with a higher percentage of silt and clay, or a better term, a sand with a *soil* component that provides better soil properties yet still maintains good drainage. This finer sand is more stable to compact, is usually much less expensive than a sand approved for greens construction, and is far less droughty in nature. This type sand is more like a mason-type, river sand. Figure 3.11 shows a prime example of sand being used for tee capping. Figure 3.12 shows the proper use of grade with the underlying soil subgrade as compared to the finish grade. Some golf contractors install a 3% slope to the subgrade and then bring the tee surface level with the sand cap while others include the 3% grade on the subbase but then use a 1% to 1.5% grade with the sand cap sloped in the same direction as the subgrade. Depending on the surface area of the particular tee box, the contractor should be allowed to use the method of construction with which they have had the best experience. Most importantly, the tees' surface and subsurface

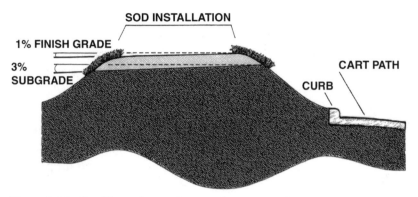

Figure 3.12. *Possible sand capped tee grading cross section.*

drainage slope must be carried away from the golfing traffic. The tee box should either be sloped to the side, or sloped to the back and to the side *away* from the cart path.

Using a too-coarse sand for tees makes it very difficult to establish firmness in the sand and also creates a very droughty root zone media. Moreover, very wet conditions result on the lower drainage side of the tee because more water is sheet drained at the soil/sand interface onto the slope. Fine sands are definitely the key to successful sand capped tee construction, preferably using a sand/amendment mix.

Proper green and tee mix compaction during the final stages of construction is a very important aspect of quality construction. At worst, many times greens are planted on a soft and unstable mix, creating a situation where surface contours shift, changing original architect design. At best, the resulting rough surfaces will require tremendous amounts of vertical mowing, topdressing, rolling, and aerifying to produce smooth playing surfaces. All these extreme concerns can be eliminated through proper greens mix compaction, firming and smoothing prior to planting. Otherwise, a course can experience one to two years of intense cultural programs before absolute smoothness is achieved.

Ideally, the greens mix should be in place and spread at least six weeks before planting. This allows settling time from rain and irrigation cycling. Irrigation cycling is critical to ensure proper operation of the irrigation system and to completely firm the greens mix prior to planting. Adding moisture in the mix is the only way to produce firmness and stability. Once the compaction is properly managed the final float of the putting surface prior to seeding will ensure that surface contours remain as designed and built. More discussion on this later.

The same is true with tees capped with sand for improved playing and growing conditions. This sand must be properly firmed with rolling and compaction through irrigation/rain prior to the final float. If the tee mix or the green mix still seems too unstable at planting time, then heavy irrigation followed by rolling will increase compaction prior to the final floating. Unstable surfaces should never be planted. The surface is too soft if it ruts more than .5 inches when applying preplants with a push-type rotary spreader containing 35–40 lb. of material.

Laser leveling tees has become an industry standard for new construction or renovation. It provides firm, true surfaces that can be set to any grade, angle, or pitch. Laser leveling maximizes surface drainage while producing a virtually level playing surface with percent grade and pitch direction set according to specifications. Uneven tees are a pet peeve of all golfers.

Laser leveling can match a lesser finish grade with a greater subgrade when utilizing the sand capped tee construction outlined above. This process produces excellent firmness in a well-selected sand or sand mix on the tee surfaces.

The irrigation trenches surrounding green complexes are the most common problem areas of trench settling. Trenches cut on angles or bends are more difficult to compact than long straight runs such as in the fairway. Consequently, a quality irrigation contractor will hand tamp these trenches to make sure compaction is complete. Also, running the irrigation system prior to final shaping or possibly hand watering the trenches from the green's quick coupler will help ensure settling as much as possible. Remember, there must be adequate soil moisture before compaction can be efficient (Figure 3.13).

Another QC item is compaction of soil at cart path edges and behind curbs. Quality golf course contractors take great

Figure 3.13. *The result of poor trench compaction—very labor-intensive to repair.*

pains to make sure that soil is properly placed and compacted along cart path edges and behind curbs to ensure the stability of the path edges. If compaction is inadequate, cart path edges can break off and curbs can be moved because of the lack of soil support. Soil settling from improper compaction creates another nightmare of labor-intensive, hard work during the late grow-in or in the first year of maintenance. Figure 3.14 is a good example of proper compaction and smoothing along cart path edges prior to planting.

Bunker Management

When is the correct time for bunker sand installation during the construction process? Many owners and superintendents put sand into the bunkers after grassing as part of the grow-in operation. This is a mistake as it creates tremendous damage and extra work

Figure 3.14. Cart path edge filling and compaction.

for the grow-in superintendent, and greatly increases the cost of bunker construction.

Bunker sand is best put into the bunkers before grassing begins because truck scars from hauling sand into the bunkers can easily be removed. Installing sand piles into bunkers prior to planting also maintains the integrity of the drainlines by protecting them from silt. Contamination of the sand piled in the bunkers should be of little concern since siltation damage from rain splashing onto the bottom of the bunker and splashing soil up on the piles of sand is inconsequential. Soil is only splashed up 10″ to 12″ onto the pile and it can easily be skimmed off the base of a sand pile with a shovel. More important is the added benefit of sand piled onto the drain field to keep it completely protected from siltation during grow-in.

If golf course construction occurs in a very windy climate, sand piles may be covered with plastic to protect them from wind erosion. In any circumstance, putting sand in the bunkers prior to planting is the proper sequence of bunker construction.

The damage of waiting to install sand in the bunkers after

Figure 3.15. Getting sand to the bunkers is the most difficult job.

grassing is vastly underestimated during grow-in and initial establishment. Figure 3.15 shows the magnitude of this tedious repair operation, and Figure 3.16 identifies the damage that occurs from getting sand into the bunkers, even with plywood protection put down to better carry the traffic. It is important to realize that all the areas where this heavy traffic will be concentrated is establishing turf, is very soft, and has yet to develop any degree of firmness for heavier traffic. Furthermore, larger hauling equipment cannot be used, so the time and expense alone of installing the sand after grassing is greatly increased.

Do not overlook this seemingly small task in the grow-in and establishment phase of the operation. Out of sequence, this is a very costly, labor-intensive operation. Only bunker sand spreading, edge trimming, and cleanup should be needed after grow-in to finish the bunkers (Figure 3.17).

Liners are always a question when building or renovating bunkers. Some superintendents have had excellent success with liners while others had very poor experiences. The two main problem points with liners have historically been:

Figure 3.16. Bunker sand installation during grow-in produces tremendous repair work.

Figure 3.17. Bunker sand is installed before planting.

64

1. anchoring them in place, and
2. snagging them with the mechanical rake or the bunker edger

The type of material utilized, method of anchoring and securing at the edges, and their use in conjunction with the drainage system should be carefully weighed. Seek advice from other superintendents who have utilized liners and qualified builders who have successfully installed them.

One of the most successful uses of liners is preventing the migration of small gravel or shale upward in the sand by bunker raking. Some soils have significant amounts of shale or gravel in their composition and this quickly becomes a problem for consistency and in the blasting of the gravel onto the putting green, from bunker shots.

Another bunker construction method, detailed in Figure 3.18, has been used to reduce the washing down of sand on slopes. However, this procedure has not been effective on bunker designs with sand flashed high on the slopes because the sand would be so deep at the slope base adequate firming would be virtually impossible. Figure 3.18 details this method.

When the wall/floor union is cut at 90 degrees, the water does not wash the sand down as it sheet drains along the subgrade. This

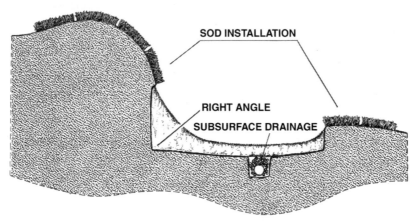

Figure 3.18. *Bunker construction to reduce slope washing.*

65

method allows the water to be moved through the sand and channeled in the drain tiles without sand movement on the slope. Again, this method is most effective on lower flashed sand faces.

Seedbed Preparation

Stripping and stockpiling topsoil, later to be capped over various golf course features, should be a part of any construction project. Unfortunately some golf courses are built today on sites barren of soil that would be classified as topsoil. Topsoil must also be tested to determine its nutritional value. If availability is limited, to only go on green and tee slopes, then the preplant needs of these slope areas will probably be drastically different than on other areas of the course. Testing of all soil types also carries over to the soil classifications on the course, which is detailed in Chapter 4.

Regardless of the type of soil predominant to the area or the golf course contractor doing the work, there is a certain amount of hand work to be done in the final prep prior to seeding. Hand raking green perimeter areas, collars, tee tops and tee shoulders are all a must in quality seedbed preparation. It is critical to always focus on the importance of seedbed preparation prior to planting as it is the last time to significantly change or improve the quality of the surface before use of a combination of aerifiers, topdressers, and rollers. These cultural programs are designed to "fine-tune" a surface rather than to smooth poor seedbed preparation, and that is why detailed seedbed preparation is so critical. This is particularly a QC job of the superintendent in the final stages of construction, especially with nongolf course builders. On some areas of the golf course, it will be impossible to effectively prepare a fine seedbed mechanically, and hand raking is crucial. Moreover, a degree of hand work is needed on the fairways for the final rock and stick removal prior to planting. This is not a difficult or time-consuming job when quality construction has been the norm throughout the project.

Sites that are particularly rocky should have a type of mechanical rock remover operated over the surfaces so the remaining rock removal by hand is simply a matter of detail instead of a large operation. Again, the quality of golf course construction is reflected in these details prior to planting. The superintendent is the person that will live with trying to make quality playing surfaces out of poor seedbed preparation if tight QC is not maintained during this operation. As with preplant fertilizers, once the grass is planted, the ability to make drastic changes or modifications is eliminated and surface cleanup becomes a significant task.

Seedbed preparation should be exactly the same for sod as for seed. Oftentimes, the same attention to detail is not given to sod prep, prohibiting aggressive rooting. The top two to three inches must be loose for sod rooting, just as it must be for seedling root development.

The other key for the best sodding results is that tightly laying the sod rolls minimizes air spaces between rolls or pieces, preventing drying out at the edges, and consequently much better health. The same is true under sod rolls. Rolling after installation removes air pockets, thus increasing surface smoothness and promoting root development. This process should not be left out on any sodded area.

Haul Roads

Haul roads are necessary in any construction/renovation project and this traffic must be planned for and managed. Location and compaction elimination within the haul roads prior to seedbed preparation are two primary considerations. Figure 3.19 is an example of a haul road during the construction phase, a necessary part of any construction activity.

Some golf course projects experience random traffic flow over the golf course instead of having concentrated haul roads for traffic movement (Figure 3.20). It is best to concentrate all traffic on spe-

Figure 3.19. Haul roads are part of any construction.

Figure 3.20. Traffic should be confined to reduce compaction.

cific haul roads because these smaller compaction areas can be corrected more effectively and efficiently.

Ideally, the haul roads or traffic roads should be the cart path beds or trails when possible, because this is the one area on the golf course where compaction is welcomed for an improved cart path bed. Although usually not possible throughout the entire golf course, this is possible on a good portion and can be planned for with proper construction management from the superintendent's perspective. A quality golf course builder can assist in this planning without problems. Their only restriction is that traffic flow areas not be restrictive to large equipment.

When final shaping is complete, it is time to get rid of the compaction in the haul roads. This is done by plowing the haul roads prior to final seedbed preparation, which eliminates the deep compaction from continued traffic and allows for uniform turfgrass growth throughout the area. Without plowing, the haul road can be quickly masked over with seedbed preparation and floating of the fairway. These areas would not be evident again until turf establishment is begun. Since the hard pan was not removed by plowing, this area will always be a problem for quality turfgrass growth because the compaction still remains. Identifying haul road locations on field plans and then getting rid of their compaction is part of the initial punch list and is a vital quality control issue for the superintendent (Figure 3.21).

Drainage

Every superintendent knows that drainage is one of the most critical issues of golf course construction or renovation, emphasizing the old adage that the three most important concerns for golf course construction are drainage, drainage, and drainage. A quality irrigation system allows for the addition of more water on an as-needed basis, but poor drainage design, surface or subsurface, prevents water removal from the golf course. Let's consider some key

69

Figure 3.21. *Haul roads require plowing before final floating.*

drainage issues during the construction/renovation process and key quality control issues that the owner rep must be aware of.

It is necessary for the superintendent to monitor grading and firmness of the area around the catch basins. Quality golf course contractors thoroughly compact and grade, usually with a box blade, around basin areas during final grading to establish smooth and continuous fall over the basin area. Grades must also be set to the proper degree of slope to ensure surface drainage, as an excessive grade will encourage too rapid a water movement. Remember that soil erosion is enhanced as much by the velocity of water as the volume. This is why the percent slope for surface drainage has some specific parameters. Engineers suggest a minimum slope of 1.5% to move water on the surface in an established turf area. However, a surface slope of 2.5% or slightly greater is considered more efficient for water movement.

Ideally the basin is sodded with a 2- to 3-foot band around the drain grate to help ensure the integrity of the grade. This also greatly reduces the tendency for erosion damage around a basin, which is further explained in Chapter 7. The detail to finish grade

Figure 3.22. Catch basin grade and firmness are critical.

work around drainage basins cannot be overemphasized because if proper grades are not established or pockets are left in these basins, they will continually hold water until the superintendent removes the sod, reestablishes grade, and then replaces the sod. This costly extra job is avoidable with good construction technique and proper QC monitoring.

Figure 3.22 shows the ideal catch basin installation method for golf courses. Properly designed and graded catch basins quickly remove channeled storm water off the surface. However, the soil immediately around a catch basin tends to be saturated for a longer period of time in heavier soils because the concentration of water around the catch basin can only be removed when it is *surface* water. Once the soil is saturated around the basin, it can remain quite wet and soft for days and is subject to rutting from mowing or golf cart traffic. Installing a gravel jacket around the catch basin and the basin riser is an effective way to deal with this problem. A short section of the riser pipe from the bottom of the basin into the storm drain system is perforated rather than solid. This allows the concentrated water in the soil profile around the basin to be quickly removed through the gravel, into the perforated tile, and finally into the storm drainage. This gravel jacket is only 4 to 6 inches thick and usually 10 to 12 inches deep because a greater gravel volume is not needed for the removal of this soil-water concentration. Appendix 11 reflects this construction technique.

Note that "knock-outs" should be included in large catch basin boxes. If boxes are site-built to accommodate varying sizes of pipe or to reduce some of the cost from prefab catch basin boxes, then a knock-out should be built into the box on all sides. This is simply a thin-walled area in the catch basin wall construction which is easily punched through if additional french drainpipe needs to be added into an area in the future. Having an outfall for french drain installations is often one of the most difficult parts of installation. Whenever a large catch basin is installed it should have easy accessibility into the box with drainage pipe additions. Many prefab boxes have knock-out points already included in their sides, but this should be a QC evaluation done by the owner rep.

Flushouts are an important part of any drain system and were virtually unused in the golf course construction business until several years ago. Flushouts are an access to the high end of the drain system, so high pressure water can be introduced on a regular basis to keep the drainlines free.

Flushouts consist of extending the end of the mainline, which is usually buried below ground, out to the edge of the green or bunker and then angled up to above ground. Figure 3.23 shows a common example of the most effective way of identifying the flushout location. Usually large metal washers are installed on the drain tile cap with a self-tapping bolt. This flushout location is then marked on the irrigation asbuilt, measured off of two irrigation heads. It is then buried below grade and can easily be found with a metal detector. The flushouts are sometimes put into small valve boxes in the green perimeter. This is not the preferred method of installation since their accessibility is very infrequent and their location is usually the outer portions of the collar, so an additional valve box here is highly exposed to golf traffic and not recommended. Bunker flushouts can be more difficult to relocate because of a lack of cross-measurement references from adjacent irrigation

Figure 3.23. *Flushouts are inexpensive and a major component of drainage.*

heads. Therefore, care must be taken to ensure their location on an asbuilt and then be properly marked with metal washers so they can be found with a metal detector.

Flushouts require careful location identification for servicing, but what about drainline location? Many contractors install a #14 gauge irrigation wire in the trench with the pipe to serve as a tracking wire. The end is accessed in a valve box or at the flushout location and the pipe location is easily identified with a wire locator. Green drainage outfalls, bunker drainlines, french drains, fairway tiling, and even the green core have been effectively marked by use of tracking wire installed in the trench with the pipe.

Every golf course superintendent knows that eventually a bunker drain becomes clogged and will need unplugging or replacing. As green complexes are constructed, sometimes the green side bunker drains have been tied into the green drainage to save pipe and trenching. This is an absolute "no" in drainage technique. Eventually this will cause plugging in the green's drainage if bunkers are piped through the green. Bunkers are often tied to the green drain outfall pipe, and a common pipe drains both green and bunker when a long outfall run is necessary. This common exit pipe is usually upgraded to a six-inch diameter to accommodate both drainage systems. Annual flushing can keep the mainline free-flowing but bunkers drained through a green will create problems in the future. Appendix 12 diagrams the proper design.

Another type of golf course drainage, open stone drains, are very effective for removing surface water or drying up an excessively wet subsurface drainage area. There are limitations to open stone drains in the golf course setting because they are not conducive to a golfer taking a divot on top of one. When open drains have been installed and the grass creeps back over the surface, they are virtually invisible. If a golfer plays a shot from on top of the stone drain, they can take a divot into the gravel and damage their club and/or their wrist. Consequently, open stone drains should not be used in an area subject to play. However, if an area is out-of-play and the open stone drain is deemed appropriate, then their installation is feasible (Figure 3.24).

Sometimes too large a stone is used for proper open stone drains. These drains work best over time using a pea-gravel size stone or stone under 0.5 inch. Drains built with larger .75 to 1 inch stone and then capping the larger stone with 2 to 3 inches of pea gravel to bring it level with the surface are some of the best.

Oftentimes low areas on the golf course cannot be gravity drained with a french drain or open stone drain because of higher elevations on all sides. These areas should be drained with a siphon system which would be required if such an area received a high volume of water from storm runoff. These system details and uses can be obtained from a drainage engineer.

For low areas that collect smaller amounts of water but still need some surface drainage assistance, a sump well system is effective. Sump wells are simply a vertical hole bored as deeply as possible and filled with drainage stone. These have been effectively used to provide surface drainage relief in areas that cannot be efficiently french drained because of topography and/or location.

Normally sump wells are installed with a tractor-mounted auger with the holes drilled from 6″ to 12″ in diameter to a depth ideally of 4′ to 5′. These sump wells can collect surface water and

Figure 3.24. *Careful location of open stone drains is important.*

provide greater dispersion of that water into the soil profile while relieving the standing surface water rapidly. Sump wells are inexpensive and can often alleviate a problem that would otherwise be very expensive to fix.

Head walls, the concrete wall surrounding a pipe end, are

Figure 3.25. *Head walls are not just for improved aesthetics.*

often considered as only for aesthetics. This may be true in cases on the outfall end of the pipe, but they play an important role on the intake end of storm drainage pipe. Head walls serve to capture surface water more efficiently and channel it into the pipe (Figure 3.25)

Just as importantly, they reduce erosion around the pipe inlet that occurs in heavy rains. When a high volume of water is forced into storm drains, a swirling action often occurs, creating erosion around the pipe. This can lead to water "blowing out" a pipe, where the water force along the outside of the pipe erodes the loose soil in the trench that was compacted around the pipe.

Management of surface water movement on the golf course is critical not only for storm water removal, but agronomically as well. Surface water from green or tee perimeters should never wash across the green or tee surface. Figure 3.26 is an example of a severe green perimeter slope that is contoured directly onto the putting surface. In this example, all agronomic programs (pesticide and chemical applications) performed in the green perimeter areas will affect the putting surface by water movement. Runoff directed

Figure 3.26. *Perimeter slope directly onto putting surface.*

onto greens is not just agronomic in nature, but is a concern for construction too. Figure 3.27 shows how silt movement onto a putting surface can be prevented. If this damage occurs, it can create problems for the future of that green. Figure 3.28 depicts the silt deposit and crusting damage that occurred on greens mix from an adjacent slope lacking a drainage swale prior to reaching the putting surface. These areas in golf course design, though not ideal from a drainage standpoint, are sometimes essential from a design standpoint. Learn to recognize these areas and protect them during construction and grow-in. Management in the green perimeter should also be correlated to accommodate these special concern areas.

In Figure 3.29 note that the sediment which moved onto the greens mix has not yet created significant damage because it is still on the surface and has not migrated into the soil profile. Future rain or irrigation cycles will create long-term damage, because now that the contamination is on the greens mix, rain will work it into the soil profile. This is a QC job for the owner rep, because when the initial damage occurs it should be skimmed off the top imme-

Figure 3.27. *Protection is required for siltation from perimeter slopes.*

Figure 3.28. *Silt deposit left after rainstorm.*

Figure 3.29. *Contamination is at the surface until subsequent rains.*

diately to prevent long-term damage, and the contamination source corrected. Delaying this silt removal can create long-term consequences that Figure 3.30, Plate 2, well illustrates.

Siltation damage to drainage systems can have permanent consequences as seen by Figure 3.31. Water is being held in a straight sand green profile directly over an existing drainline covered with pea gravel. When the core cavity was exposed and the herringbone system installed, a rainstorm washed silt onto the green core. Only the upper 2″ of gravel was skimmed, replaced, and then the greens profile sand was put into place. When grow-in irrigation began, this green quickly held water. This is because the slotted drainlines were completely encapsulated with a silt band and the gravel immediately surrounding the pipe was silt contaminated. Removing the top 2" of gravel above the pipe only solved a small portion of the problem.

The scenario would be the same over the gravel blanket of a green or over bunker drains as identified in Figures 3-32 and 3-33. If siltation of a drainline occurs after installation, it is critical that the entire gravel profile in the trench be removed with the pipe, the

Figure 3.31. Silt contaminated drainlines before mix installation—now clogged.

pipe cleaned so the slits are clear, and then the gravel and tile replaced. It is crucial to protect drainlines from damage once they are installed and to closely monitor drain tile and gravel cleanliness until the profile is permanently put into place.

Storm water removal from tee and green perimeters is very important, but surface drainage patterns are equally important around bunkers. Figure 3.34 shows an example of surface water that moves directly through this bunker from poor perimeter surface drainage design. Consequently, the bunker drainage is contaminated only six months after opening. Poor perimeter surface drainage can increase this maintenance expense two or three times because of the continual need to replace sand that washed from rain or irrigation. Quality golf course contractors ensure that proper drainage swells are installed around bunkers to prevent this damage. However, it should be the duty of the owner rep to double-check this quality control area to prevent future turf management problems. The benefits of a simple surveying level and rod are immeasurable with these types of quality control checks. Ob-

Figure 3.32. *Silt contamination does not remain just at surface on the gravel layer.*

Figure 3.33. *Bunker drainlines also require siltation protection.*

Figure 3.34. *Surface water should never move through bunkers.*

serving surface water flow after final grading is also very effective in identifying such problems.

Tree Concerns

Tree care during and after construction is a major concern because the desire to keep specimen trees is usually shared by all developers. It is commonly thought that if a tree is not struck by a bulldozer during construction or if the tree is not backfilled significantly then no harm is done. However, few recognize the delicacy of the tree root system and what factors cause the greatest detriment to its health.

The best way to protect key trees is to rope off the entire root system (drip line) around the tree initially so it escapes any construction activity, including traffic. Heavy traffic, regardless of the type of equipment, is the biggest detriment to tree health during construction and where most trees are exposed to damage. Protec-

Figure 3.35. Traffic on tree root systems is very detrimental.

tive fencing should go outside the drip line by about 10% to provide the greatest protection to the root system. Figure 3.35 is an example of problems with tree protection.

Light construction activity, be it slight soil removal of a few inches in depth around the tree or significant plowing of the surface prior to planting grass within the drip line, is also hazardous. Arborists agree that greater than 80% of a tree's root system is in the upper one foot of soil, and partial removal of the topsoil layer or significant plowing is tremendously destructive to a tree's feeder root system.

Backfilling is additionally detrimental to existing trees. Backfilling buries the original grade around the tree and thus buries the active root system beyond reasonable aeration in the upper topsoil layers. Root systems of all plants must have oxygen to grow and when this soil:air:water ratio is disturbed, plant health is sacrificed. Figure 3.36 shows an example of backfilling around existing trees while protecting the original surface level and root system. Again, a local arborist can provide excellent recommendations for backfill-

Figure 3.36. Fill around trees must have careful planning.

ing around key trees if necessary and the sensitivity of that particular species to backfilling.

During construction it is beneficial to root prune key trees as early as possible so the active root system can be concentrated into an area undisturbed by construction activity. Root prune with a trencher or subsoil plow and install the fencing to delineate the boundary of protection around the tree. As mentioned above, many arborists recommend this protection area be extended about 10% beyond the drip line of the tree. However, to be sure of the root system's extensiveness, experiment with pit digs to find how far the root system is from the tree trunk. Do this on trees of the same species that are not to be saved so damage does not occur to specimen trees. Begin by digging pits out beyond the drip line to determine the depth and size of the feeder roots, then work back toward the trunk of the tree until the root mass profile and its diameter is seen. Roots larger than about 1 inch in diameter should not be cut.

A 3- to 5-inch deep layer of mulch should be placed underneath the canopy in the root pruned area which provides much improved soil moisture for damaged root systems. Ideally, maintain-

Figure 3.37. Mulch cover should be planned under trees in some areas.

ing a mulch cover rather than turfgrass under specimen tree canopies reduces competition and traffic underneath the canopy of the tree. Again, remember that traffic on a tree root system is considered by arborists to be the leading cause of tree decline and death versus actual construction activities (Figure 3.37).

Root loss will shock a tree. Consequently, selective pruning of the canopy must offset this loss. Remove dead or diseased limbs first, then additional limbs can be removed to encourage a better shape and uniformity. However, arborists recommend that no more than 20% to 25% of the canopy mass be pruned away from a tree at any one time (Figure 3.38). As a general rule, remove only limbs about 2 inches or smaller in diameter. This is an excellent time to raise canopies by pruning to eliminate problems with golfing or mowing traffic or to improve air circulation.

The survival rate of saved trees is much higher than that of transplanted trees. Arborist research states that saved trees may lose about 50% or less of the root system whereas transplanted trees lose an estimated 90% of their root system. Consequently, saved trees go through much less stress and have a much greater survival rate.

Figure 3.38. *Tree surgeons can maximize sunlight and air flow under trees.*

Protecting a group of trees versus single trees is recommended for better survivability. Often in golf course construction, single specimen trees are identified as key trees for aesthetics or playability and then construction activities are carried out around these trees. However, if groups of trees can be identified whenever possible, survival will be more successful and is oftentimes a more natural looking landscape than singled out specimen trees.

Another critical tree care area is making budgetary provisions for tree removal during the first year of construction. With the best of efforts made to protect trees during the construction process, a certain amount of attrition will occur from construction activities or direct wind exposure. Figure 3.39 is a prime example of this occurence. The first year budget should reflect a line item for specific tree removal because of attrition and a line item for tree replacement and additions. A line item for tree planting and/or replacement should be in the budget every year. This is also an excellent opportunity to improve the landscaping and aesthetic quality of the golf course and at the same time select species and placement of

Figure 3.39. *Construction attrition of some trees is expected.*

trees that will not harm the turf through morning shade and/or poor air circulation.

A final important consideration of tree care to be outlined—lightning protection. When specimen trees are suddenly in an open environment from a previously heavily wooded one, they are more subject to wind damage and lightning strikes. These trees often become the high point of the surrounding area as shown in Figure 3.40, as a feature of a golf hole, and which consequently is a "target" for a lightning strike. Arborists have very detailed specifications for properly protecting key trees with lightning protection, which should be used immediately.

Shade/Air Movement/Root Competition

Shade and air circulation are critical factors affecting turfgrass growth and health regardless of climate. All older golf courses face

87

Figure 3.40. Lightning protection is required on 'exposed' trees.

increasingly challenging problems from shade and air circulation as trees grow and mature. Any new golf courses built today in a deep valley or in heavily forested area will face the same concerns when tees or greens are tucked into very restrictive areas. What effect do these factors have on a particular area?

Shade is of major concern for turf health in golf course management. However, sometimes not shade, but air circulation is the problem. It is important to evaluate both factors to determine if one or both is the problem. Morning shade is far more detrimental to turfgrass health than afternoon shade because plants have greater growth activity in the morning hours after sunrise. Morning shade greatly reduces this growth and prevents the turf from drying out, in turn increasing disease activity. Oftentimes, trees are thinned or removed to solve a shade problem when in fact the wrong trees are removed. What is the best evaluation method to determine which trees, if any, are a problem and how does one summarize the magnitude of the problem?

It is impossible to fully evaluate the shade pattern of a particular tree or group of trees until they are observed at regular in-

Figure 3.41. *Shade and air movement evaluation can be done during construction.*

tervals from sunrise until the area is in full sun. Upon close examination of shade patterns, you often find that trees a little further from the green rather than trees closest to the green actually produce the shade cover. Observing shade patterns and movements for a few days is the proper way to determine exactly which trees create shade and to what extent. Will selective pruning make the shade problem tolerable? Will raising the tree canopies allow for early morning sun so the shade is only late morning or midday? Or does proper turf health require removal of the tree? Careful examination of shade patterns will allow for educated, corrective decisions. Remember that shade patterns are cast longer in the winter because the sun is at a lower angle in the sky. Winter shade problems, however, tend to be less of a problem than shade prevalent during the growing season. The type of tree, evergreen or deciduous, is also very important when evaluating winter shade.

Air movement is critical when evaluating a tree's effect on greens or tees. Low hanging branches or understory growth can diminish air circulation to a green. Ideally an air flow of 3 to 4 mph is desired across the turf canopy to dry the surface and reduce temperatures. The proper way to evaluate air movement and the predominant direction of seasonal air flow is from a local weather office. Then, with a compass, seasonal air flow predominance can be evaluated and areas of potential selective pruning and thinning can be better identified.

For example, a green such as in Figure 3.41 is surrounded on three sides by a heavily wooded area. However, the morning sun and the predominant direction of air movement comes directly up the fairway as you view the picture. Therefore, this green does not have a particular shade or air movement problem because of the orientation of the trees to the morning sun and predominant air flow. Directional air flow restrictions are located with the compass and then identified on the green details, field records, and asbuilts (Appendix 2). These problem areas can often be selectively thinned and air flow improved without "massive" tree removal. When evaluating these concerns, tree removal is not always the answer. Removing a tree because of its proximity to the green only to later realize that it created no problems for the green is a detriment to the environment and the aesthetics. Carefully weighing tree proximity, canopy effects on turf health, and the tree's restriction of air movement must all be simultaneously weighed for proper tree evaluation. On existing courses, removing the wrong trees eliminates credibility to then remove more trees. Careful evaluation is the key.

While determining tree removal on the golf course, it is an excellent time to evaluate whether a green site may need fans for proper air movement. If this is the case, providing power to the green complex will never be easier or less expensive than when installing irrigation. Providing power to a remote green sight may be coordinated with the irrigation designer or may be supplied from an adjacent area such as a rest station or a transformer from a cul de sac in a development golf course. Here is another example where preplanning can save tremendous headaches and expense.

The last aspect of evaluating tree effects on turfgrass areas is root competition. Tree root competition can be a tremendous problem to turfgrass health, but tree root pruning can effectively control

Figure 3.42. *Root competition can be anticipated.*

many of these problems. Consult a local arborist regarding the tree species, the intensity of the root pruning that can be done in relation to distance from the trunk, and any other special concerns. During construction, root pruning with a trencher and lining the trench wall with a plastic barrier will slow down the reinfestation of tree roots into an area. Tree root pruning should be a part of the overall tree care program during construction. Proper deep root feeding following construction must be done so roots can recover from the stunting caused by construction and root pruning. Figure 3.42 is a good example of proper root pruning technique in turfgrass areas. This could have been as easily accomplished during construction or renovation if needed.

Again, remember that shade alone is often not the problem. All three factors—shade, air flow, and root competition, must be carefully weighed *simultaneously*, because usually it is a combination of problems instead of just one. Too, selective pruning and raising canopies can solve shade and air problems many times where tree removal was the first consideration. Pruning does not always work but it should be closely evaluated and tested for a period of time prior to complete tree removal. Unnecessary tree removal is not an environmentally sound decision.

Tree Plantings

Tree selection and locations must be carefully weighed during the long-range plan of a renovation project or during the master plan of a new course. Many times tree plantings do not match up with native species indigenous to the area. Careful selection of trees to be used on the golf course requires specialized species selection and growth characteristic understanding. Improper tree placement and management is critical for turfgrass and tree health. For example, tree placement on the southeastern side of a putting green can create shade and even air movement problems as the tree matures. Likewise, once the investment is made in tree planting on the golf course, money for tree management must be included in the budget. Tree planting during grow-in is not recommended unless staff and budget specifically allow for tree care. Grow-in is unbelievably hectic and the responsibility of hand watering trees can easily be overlooked.

CHAPTER FOUR

READY TO PLANT?: PLANTING DETAILS

Preplant

Preplant fertilization is often taken for granted. Most golf course specification plans detail preplant fertilization without ever taking a soil test. Soil needs cannot be specifically determined without soil tests, and there will never be a better opportunity to correct basic soil nutrition needs than before planting.

Preplant specifications in the budget should be generous. The more expensive preplant programs specified by many designers today have adequate money for most any site-specific soil amendments and nutritional requirements. Materials, analyses, and even rates are specified. Architect preplant recommendations, however, should be used for budgetary purposes only and not for actual soil needs (Figure 4.1).

Unfortunately, many turf managers apply preplant materials according to preplant recommendations in the golf course development specifications instead of taking detailed soil tests to discern

Figure 4.1. *Preplant application on the golf course.*

<u>actual</u> needs. Applying soil amendments is not enough. For example, applying the wrong liming material can put the soil's nutritional balance in a worse condition than if no lime had been applied. Why even consider actually applying preplant specifications not based on soil tests? (Figure 4.2).

As mentioned, there will never be a better time to correct soil nutritional needs than before the grass is planted. The amounts and types of materials that can be applied and worked into the upper portion of the soil is almost limitless if soil conditions dictate drastic measures. The golf course plan (in some cases) or the erosion and sediment control plan is an excellent source of determining the various soil types on different areas of the course. If small amounts of cut and fill are being done, then soil tests can be taken most any time during the construction phase. However, if large amounts of cut and fill are part of the design/build process, then rough shaping should be done before soil testing. This ensures that the upper surface of soil that is now to be the "topsoil" is what is being tested throughout the course.

When evaluating soil tests, key values must be closely

Figure 4.2. *Preplant materials on greens.*

weighed to determine the true nutritional balance of a soil. Pounds-per-acre categories only tell nutrient volume, not the availability of nutrients to plants. The percent base saturation (%BS) is the key to determining nutritional proportions and availability.

Base saturation is the contribution of each cation to the total holding capacity or base saturation of a soil. Percentages of the key cations as well as the ratios with others have a drastic effect on nutrient availability. Table 4.1 outlines target base saturation percentages of the cations in a properly balanced soil. Note that the Ca:Mg ratio is different for soils and sands. Sands are more deficient in magnesium than are soils and so require a target range of higher magnesium levels. Calcium and magnesium ideally should add up to 80% of the BS—60:20 in sands and 70:10 in soils.

If a soil test is evaluated with an appreciation for %BS categories, the turf manager is better able to determine the needs of the soil and make more accurate corrections. For example, if the soil test indicates very low calcium but the magnesium is close to adequate, then calcitic lime is needed, not dolomitic lime. Carefully weighing the Ca:Mg ratios helps the turf manager to select the best

Table 4.1. Proper Percent Base Saturation Target Values

	Target % Base Saturations	
	SAND	SOIL
Ca	60%	70%
Mg	20%	10%
	KEY: 80% Combined	
H	10–15%	10–15%
K	3–5%	3–5%
Na	2–4%	2–4%
Other Bases	2–4%	2–4%

liming material for a soil's needs. Applying dolomitic lime for calcium to a soil that already has a greater than desired magnesium level will further complicate the nutrient imbalance. However, usually architect specifications recommend dolomitic lime, even with a rate per acre to soils, without a soil test. Don't trust a prewritten golf course construction spec to know preplant needs. That is the job of the turf manager and no one can do soil evaluation better than the turf manager at his/her site when they have pulled their own soil samples representative of the various soil types on the course.

Phosphorous has been shown to promote lateral root growth in many plants, thus illustrating the necessity of generous amounts of phosphorous availability during establishment. Phosphorous is the foundation nutrient in starter fertilizers because of this need. Phosphorus is immobile in the soil, and greater soil solution concentrations are needed at seeding. Appendix 13 better illustrates this.

Liming is the most basic of soil amendment applications. Most lime recommendations assume that the turf manager will use materials which have the same neutralizing potential as pure calcium carbonate.

This means that if a soil test recommends 40 lb. of limestone per 1,000 sq. ft., use a lime source which raises the soil pH to the same degree as 40 lb. of pure calcium carbonate would do at the same rate. A liming material with the same neutralizing potential as calcium carbonate has a calcium carbonate equivalent (CCE) of

100%. Turf managers should identify the CCE of the liming material to be used and adjust rates accordingly to match a 100% CCE. Soil tests will indicate what the analysis shows, but most turf managers are not aware of this fact and consequently might underlime according to need. As mentioned before, there will never be a better time to make soil adjustments than before planting. Details like this are critical to create the best growing environment possible.

Particle size distribution of lime is another key aspect relating to its breakdown and behavior in the soil. A coarser material is slower acting than a finer material just by the physical constraints of breakdown within the soil. The best liming materials are finely and uniformly ground. A finely ground limestone contains a greater number of particles per given volume than a coarser ground material and will more rapidly affect soil pH. Limestone specifications should conform to the following sieve sizes:

- 95% of the limestone must pass through a 20 mesh per inch screen
- 60% of the limestone must pass through a 60 mesh per inch screen
- 50% of the limestone must pass through a 100 mesh per inch screen.

The following formula is required when calculating the CCE of a chosen lime material:

$$\frac{\text{The lab recommended liming rate (lb. of lime per 1,000 sq. ft.)} \times 100}{\text{The CCE of the liming material to be used}}$$

This formula is used to calculate the actual rate of application needed for a liming material to provide the same effects in the soil that pure calcium carbonate would produce.

Sulfur is another nutrient not to be overlooked for its importance in soil health and turf management. Elemental sulfur (greater than 90% sulfur) is the most common and readily available source of sulfur for lowering soil pH. Many superintendents use ammonium sulfate to lower the pH of high-sand putting greens, but this practice has limited potential because of the small amounts of sulfur actually added per application and even within the course of a year. Many superintendents have steered away from sulfur coated urea (SCU) because of its potential to create black layer, but the amounts of sulfur from sulfur-coated or poly-coated fertilizers is basically insignificant and has not caused problems when sulfur from SCU fertilizers alone is used.

Sandy soils have a reduced buffering capacity when compared to heavier soils and therefore require less lime or sulfur to make pH adjustments. Sulfur applications, as with lime, must be carefully weighed according to soil tests to achieve the targeted beneficial soil adjustments. The *Western Fertilizer Handbook* (26) can supply more information on the sulfur rates necessary to lower soil pH.

When soil testing to determine initial sulfur needs or regular monitoring to determine additional needs, some very specific soil test techniques are required because of the immobility of sulfur. Soil tests should be taken as a surface test which is only about 1 inch deep. Then another soil sample is taken from the 1 inch depth down to 3 inches. This surface soil test is important because most of the sulfur applied to established turf and a large proportion applied to establishing turf will remain near the surface.

A slowly available nitrogen source should be included in preplant applications in addition to regularly used preplants such as starter fertilizer. The 2–3 weeks after seeding/sprigging requires the greatest irrigation frequencies, thus the wettest soil conditions. Making a granular application 8–12 days after planting can create significant rutting damage from application equipment. An application of slow release N at about 1.5 lb. /1000 sq. ft. usually provides the N requirements of new seedlings during germination. This N availability also promotes faster, more aggressive growth of seedlings and newly rooting sprigs. Fertigation can assist in this problem of N applications, discussed in Chapter 5. Applying slow release N at planting is another example of the benefits of planning ahead.

Matching Turf Selections to Conditions/Use

We are very fortunate today to have so many varietal choices in all turfgrass species. This flexibility allows consideration of specific features such as: color, texture, mowing height, endophyte content, drought tolerance, disease resistance, and even mowability. The best source to evaluate varietal performance across the country is the *National Turfgrass Evaluation Program* (NTEP). Subscriptions can be obtained from the United States Department of Agriculture in Beltsville, Maryland, and updated varietal performance can be examined at various locations across the country.

Grass selection in golf course development does not receive the attention it should in many respects. Researching varietal capabilities and compatibilities to the desired use and climate is not always carried out with proper detail. There is often a big concern for selecting the mix components for natural, native, or wildflower areas, but less attention is given to a mix or blend for the roughs. A blend, by definition, is more than one variety of grass of the same species blended together for improved performance. Examples of a blend would be multiple bentgrasses blended for fairways or any of the three ryegrass variety blends so commonly used in fairways and for southern overseeding. A mixture, however, is a combination of two or more different species of grass to provide improved characteristics over one species alone. An example of a mixture would be the combined use of bluegrass and ryegrass in fairways and roughs for improved performance or a combination of different fine fescues to obtain different appearance characteristics. These characteristics may include growing height, color, or even drought resistance in deep roughs.

When making varietal selections for the golf course one must lay the groundwork for making selections based on the goals of the playability of each area of the course. Factors to consider are: (1) amounts of play, (2) the desired maintenance level, (3) the aesthetic look desired by the architect, (4) the cutting height and mowing equipment used to maintain different areas, (5) the adaptability of a "suggested" variety to the environmental conditions, and (6) water quality.

Oftentimes, grass varieties are recommended by sources unqualified to make species recommendations. For example, bluegrass fairways would not meet the needs of a private course in the upper mid-Atlantic that wished to have the fairways mowed at .375 of an inch. The same might be true for recommending bentgrass greens in south Georgia. Agronomic inputs are essential at this stage of golf development and reputable architects and developers welcome the input of the golf course superintendent. Qualified architects know it is critical that selected grasses can be maintained properly to achieve the playing conditions and the aesthetic conditions they had in mind with the design characteristics.

Qualified agronomists can offer suggestions and explain limitations of species and varietal selections. Quality seed companies also have well-trained specialists who provide advice for varietal selections. Local extension specialists at state universities can offer tremendous insight in mixed component selections for native, natural, or wildflower areas. Many regional blends available today are well adapted for nearly all environments.

When making varietal selections, the question of "can I buy a sod grown with the same mixture or blend as the seeded areas?" can arise. Sod companies are aware of this need and preplanning can allow a sod company to grow sod that is well adapted or even identical to the seeded selections of chosen mixtures or blends. This "contract growing," by far the best means of obtaining uniformity, can often be coordinated through the golf course builder and the sod company by preplanning.

If this preplanning step was not incorporated, then there will be some differences in sod selection and the seeded areas. To bring consistency and uniformity over the entire area, first buy the sod that most closely matches up with the seeded varietal selection, whether it is a mix, blend, or monostand. After establishment, the sodded areas can be interseeded with the mixture or blend used in all of the areas established from seed. This will introduce the predominate grass selection into the sodded areas and bring better

Figure 4.3. *Turf-type tall fescue planned in deep roughs.*

consistency and uniformity of the seeded areas into the somewhat different sodded areas.

The seeding operation must be done with certified seed. Blue tag seed should be required for putting surfaces, and most choose blue tag seed for tee tops when bentgrass tees are selected. Similarly, certified sprigs are needed when establishing a warm season turf to ensure purity and turf true to cultivar.

Turf-type tall fescues for use in roughs have regained significant popularity over the last two to three years for two major reasons: (1) the improved density and growth habit of the grass and (2) their drought tolerance in natural rough areas and their color and texture contrast (Figure 4.3).

Turf-type tall fescues perform excellently as a deep rough turf, but the biggest problem plaguing turf-type tall fescues is being mowed too low. Turf-type tall fescues must be maintained at 2.5 to 3 inches for an ideal growth environment. Oftentimes, deep roughs are established in turf-type tall fescues and then mowed an ideal 2.75 to 3 inches. If pressure from the golfing clientele requires the superintendent to lower these cutting heights down to 2 inches or below, the turf-type tall fescues begin to significantly thin out. Pre-planning the golf course and carefully selecting varieties based on performance, climatic conditions, and the designed look and features are critical. Often turf-type tall fescues are blamed for thinning out when mowed at 2 inches or below, whereas the problem is using a grass variety out of its adaptability.

Another option when considering turf-type tall fescues is to add about 10% Kentucky bluegrass as a mixture. The bluegrass provides additional density characteristics to the mixture. This practice is very successful and popular in the mid-Atlantic and north, but the bluegrass has not performed as well in the upper parts of the southeast where turf-type tall fescues perform well by themselves. The addition of bluegrass does not allow for a lower cutting height to appease the golfing membership, but it does provide some additional density in the lower ranges of the cutting height—about 2.5 inches.

Fine fescues have also developed into a significant market in the golf course industry. They are chosen for a great number of uses, their advantages being IPM characteristics such as drought tolerance, low maintenance, texture and color differences, and their performance as natural or unmowed areas. The species of fine fescues include creeping red fescue, chewings fescue, hard fescue, and sheep fescue. As with turf-type tall fescues, the cutting height should ideally be at 2.5 inches or above but the fine leaf fescues offer a tremendous range of cutting heights from 2.5 inches to unmowed.

Fine fescues have a wide range of environmental adaptations and again the NTEP trials are the best information source when considering varietal performance. Major seed manufacturers also have excellent product information which outlines variety and species performance as related to environmental adaptability (Figure 4.4).

Research shows that hard fescue and sheep fescue are more heat and full sun tolerant than chewings or creeping red. Fertilizer applications should be low with all fine leaf fescues. Cautiously

Figure 4.4. *Fine fescue deep roughs for links style definition.*

apply herbicide as fine fescues are more sensitive to herbicide damage than are turf-type tall fescues.

Another mistaken performance quality of the fine fescues is their adaptability to deep shade. They, like most turfgrasses, have problems with *deep* shade but thrive in moderate shade. The definition of "deep shade" varies when determining the percent shade cover for turf performance. Even fine fescues have a limit to their shade tolerance.

Fine fescues are less tolerant to spring seeding than turf-type tall fescues. Spring seeding of any fescue is more difficult because of its inability to mature before the summer stress period. However, turf-type tall fescues have been shown to tolerate some early spring seeding and even some early summer seeding quite well (Table 4.2). Summer seeding of fine fescues, however, should be avoided altogether.

A mixture of the different types of fine fescues can be custom blended and site-specific, based on the look and performance desired. There are some guidelines for uses of the different fine fescue species based on adaptability. Dernoeden's research (28) recom-

Table 4.2. Suggested Planting Temperatures

	Soil Temperatures (4 inches) °F	Air Temperatures °F
Warm Season	70–80	80–90
Cool Season	60–70	75–80

mends that creeping red fescue should not be used above 10% in any mix in sunny or dry locations. Hard fescue has a deeper green color while the sheep fescue is more bluish-green. Therefore, the selection of these can provide differences in color contrast but the dominant component of the mixture is recommended to be at 80% of the mix by weight. If a mix is to be used for an unmowed or natural area with a seed head display, research recommends 40% hard fescue by weight, 40% sheep fescue, 10% chewings fescue, and 10% creeping red fescue.

The correct seeding rates for fine fescue, whether a monostand or a mixture, is 4 to 6 lb. per 1,000 sq. ft. The higher rate is suggested for a quicker establishment. Fine fescue seed has a short viable life span, with a germination rate possibly less than 50% for seed one or more years old. Therefore, fine fescues must be used within 12 months after harvest to have their greatest viability.

Proper percentages of turfgrass species in a mixture can greatly affect a mixture's performance. For example, turf-type tall fescues and bluegrasses are often blended together to create a denser turf stand than the turf-type tall fescues alone or to create a more traffic/drought stand than bluegrass alone. The proper ratio for this mixture is 90% turf-type tall fescue and 10% Kentucky bluegrass. If the bluegrass percentage were increased to 20% or 25%, the mixture would become very irregular and patchy, because the establishment of the bluegrass would be at too great a percentage due to the number of seed per pound. Ten percent provides an adequate amount of bluegrass in relationship to the number of seed per pound and a very uniform and density-improved turfgrass stand.

Seed mixture ratios and various uses are charted in Table 4.3

Table 4.3. Cool Season Grass Blend/Mixture Suggestions

Mixture	Proper Ratio by Percent	Use Area
Bluegrass/ryegrass	80/20	Fairways, immediate roughs
Bluegrass/fine fescue/ ryegrass	60/20/20	Roughs in sun
Bluegrass/fine fescue/ ryegrass	40/40/20	Roughs mostly shade
Tall fescue/bluegrass	90/10	Roughs mostly shade
Hard fescue/chewings areas/ fescue	80/20	Low maintenance Grass bunkers

according to climate, turfgrass usage, and growing condition—sun or shade. The mixtures and blends listed in Table 4.3 are suggested guidelines for species, mixtures, and ratios based on performance, environmental conditions, and playing area.

When ordering seed for new golf course construction or for one in major renovation, the planting date must be prepared for early order to ensure availability. Seed for an upcoming late summer seeding should be booked the previous winter to guarantee availability of desired varieties. This is a very important planning aspect, as in some years there are seed shortages if production has been poor. An early confirmed order can save significant headaches during years of poor seed availability.

Planting Details

The importance of seeding rates and techniques in accurate and precise application are usually underestimated. Calibration of seed to a specific seeding rate can influence stand establishment and performance significantly. The same is true with seeding technique, as a common mistake is to actually plant seeds too deep in the seeding operation. Consequently, establishment percentages are greatly reduced (Figure 4.5).

Seeding rates were established by a specific set of research de-

Figure 4.5. Cultipacker seeding is preferred technique.

tails and are based on seed count per pound and species aggressiveness. Seed size ranges from about 200,000 seeds per lb. for tall fescue to about 6,000,000 seeds per lb. for bentgrass. Seeding rates are established as a means to provide the proper rate of seed per given area. Normally, seeding rates are targeted to provide 10–25 seed per square inch for most turfgrass varieties. Bentgrass, however, because of its exceptionally small size, targets 30–60 seed per sq. inch based on the 1/2 to 1 lb. per 1,000 square feet seeding rate.

Dr. Madison's seeding rate research done in the 1960s still proves to provide the optimum rates for turfgrass establishment. He based his research on seed count and aggressiveness and his seeding rate ranges are still ideal for the best establishment in the shortest period of time. Refer to the *Agronomy Journal* (15) to review his research and findings.

Today bentgrass seeding rates at 3 lb. per 1,000 sq. ft. are common. Initial thoughts are that establishment rates will be more rapid, but in reality, long-term effects of such high seeding rates are detrimental to the overall health of the turf stand. Disease occurrence is greater and this high seeding rate is not only more expensive over the first one to two years after establishment, but is also less environmentally responsible. Additionally, higher fertility programs will be necessary to feed the excess density of plants per area, and greater amounts of water will be needed to offset wilting. Higher than recommended seeding rates do not benefit turf establishment unless a late fall seeding is necessary and the additional seeding rates are added to offset lower soil temperatures and the greater mortality rates from soil freezing. In this respect, seeding rates should be increased about 20%, not 50% as oftentimes bentgrass seeding rates are set. Seeding rates are more than just guidelines.

Seeding at a lower than optimum seeding range may result in an extensive weed problem during establishment and a poor ability of the grass to increase in density and form a sod. This can be significantly detrimental if more perennial or persistent weeds develop because of lack of competition, the primary example being annual bluegrass (*Poa annua*). On the other hand, when seeding rates are too high, disease infestations are greatly enhanced and the excessive competition creates a very spindly growth and a very fragile root system—a great target for *Pythium* or *damping-off*.

Establishment with fungicide treated seed has improved tremendously over recent years. Researchers have documented benefits such as improved development and turfgrass maturity in stands seeded with treated seed. Fungicide treated seed reduces the occurrences of *Pythium* and *damping-off* more than spray or granular applications, and reduces application traffic on an area. Remember, established areas are excessively wet and susceptible to traffic damage the first 10–14 days after seeding.

Temperature has an incredible influence on seedling germination. For example, a greater seeding rate is usually required for cool season grasses when soil temperatures dip below 60° F. This is due to the greater mortality rate from the freezing that is soon to follow. Moreover, cool season grasses establish best in early fall because soil temperatures and day lengths are beginning to decrease as the rainfall usually increases in most parts of the country. This is an ideal environment for seedling establishment and explains why late summer seeding in preparation for early fall is the ideal planting window for cool season turfgrasses.

The effects of temperature on seed germination and establishment are definitely underestimated in our industry. Owners and developers must thoroughly understand the detriments of being forced to seed a golf course well out of the ideal seeding/establishment window. Consider some of these concerns in more detail.

Figures 4.6 and 4.7 are examples of bentgrass tees/fairways established on the same golf course. Figure 4.6 is a bentgrass tee in the northern climate that was seeded in August and had germination in five to six days. Figure 4.7 is a tee on the same golf course seeded in late November. The photograph was take 28 days after seeding and still lacked germination. The difference was the soil temperature—about 68–70°F in late September versus about 45°F in late November. Blanket covers, as discussed, will increase the longevity of soil temperatures into the fall or increase the warmth of soil temperatures in the spring and in doing so will lengthen the normal growing window.

It is not uncommon for new golf course projects to run into the late fall, providing an early winter seeding window with the belief that establishment will still be possible in the early stages of winter and the turf will "catch up" quickly in the early spring. This is rarely the case. If seed lays in a wet environment over the winter, it may almost completely rot and establishment would be nearly

Figure 4.6. Bentgrass germination in 5-7 days (foreground)—soil temperature 65–70°F.

Figure 4.7. Bentgrass germination after 28 days (foreground)—soil temperature 40–45°F.

Figure 4.8. *Turf covers can be a tremendous aid in establishment.*

nonexistent, requiring reseeding the next spring. If seedlings begin the early stages of germination in the late fall and then experience significant heavy frosts, the unprotected seedlings may have a low survival rate.

Turf covers have been utilized with great success in a number of situations. Figure 4.8 is a typical use of covers to extend the growing weather window. Covers can maintain soil temperatures when establishment is attempted out of the ideal planting window. This may be a late fall/early winter seeding or late winter seeding to take advantage of spring weather.

Heavy shade can reduce establishment rates in otherwise good conditions. Covers are extremely helpful to raise soil temperatures and compensate for lesser quality growing conditions. However, there are some precautions that must be understood when using covers. Normally, covers are left on turf areas overnight and removed on sunny days with moderate temperatures to prevent excessive heat buildup. A close check for disease is critical because of increased temperatures and humidity. A preventative fungicide program is recommended to maximize the positive benefits covers provide.

There is another important consideration when missing the seeding window. It is not agronomic in nature, but will have a significant impact on the grow-in/opening time period—management team awareness. The owner representative must advise the management team of the problems arising because of poor establishment from a missed seeding window. Management must understand that this can greatly deter the opening date of the golf course the following year if an early summer opening was missed because of the significantly reduced establishment rate. This common scenario creates problems and increases the duration of the overall grow-in process. As discussed in some detail in Chapter 2, the grow-in period is normally thought of and budgeted for a six-month period. However, if three or four months of winter are put into this grow-in duration, then the grow-in period can easily be nine to ten months with a third of that consisting of a completely "no-growth" environment. Once again, it is the owner representative's responsibility to make sure that the management team understands the consequences of significantly missing an ideal seeding window in terms of golf course establishment and the subsequent opening date.

Late spring seeding is usually best for most warm season grasses because soil temperatures must be high enough to provide an adequate environment for germination and establishment. Early spring seeding of warm season grasses is rarely as successful because competition from late germinating winter annual weeds is high and temperatures too low for germination and active growth. As a rule of thumb, until the nighttime temperature low and the daytime temperature high add up to 150°F, bermudagrass will not aggressively grow. Other warm season grasses such as zoysiagrass have a slightly lower temperature threshold. This is why it is common to see bermudagrass green-up early in the spring but lacking in lateral growth or density. This temperature relationship between day and night temperatures helps one understand how sensitive warm season turf growth is to cooler temperatures. This phenomena is carried through to seeding and establishment, and is true with establishment from sprigs.

There has always been a strong controversy about fall versus spring seeding of cool season grasses, typically bentgrass green establishment or renovation. Spring seeding is generally more difficult for establishment and maturity than is early fall due to the lack of time for the cool season grass to mature and become hardy prior to summer. This is especially true on bentgrasses in the southern climates.

Fertility management and the differences between a fall versus a spring seeding, pointed out in Chapter 5, must be clearly understood. Turf managers must be extremely careful to keep nitrogen as low as possible during spring seeding establishment, yet not inhibit foliar growth. Potassium and other nutrients must be kept at optimum levels. It is necessary to develop as aggressive a turf and root system as possible in the spring while at the same time keeping nitrogen levels minimal.

Immature, cool season turfgrass root systems are very heat sensitive and subject to high temperature stress. Water management becomes even more critical during the summer months after spring seeding. A properly balanced water management program will not only keep good moisture but will maintain proper air balance which helps to moderate soil temperatures. Overwatered greens can lack a proper soil:air:water ratio balance, creating higher than normal soil temperatures and stressing turfgrasses even further.

Seeding technique has some very specific procedures to ensure the most rapid and uniform germination. The main purpose of seeding technique is twofold: (1) uniform seed distribution, and (2) best seed-soil contact. There are many variations of seeding techniques today, and seed size affects technique and the success of one method versus another. Seeding techniques vary from simply distributing the seed on top of the ground and covering with mulch to a process of actually tracking with a dimpled tire which is usually a mechanical bunker rake.

Larger seed turfgrasses such as fescues and ryegrasses are best seeded with a drop, cultipacker seeder. These ensure uniform dis-

Color Plate 1 (Figure 3.3.) On-site mixing is not recommended.

Color Plate 2 (Figure 3.30.) Siltation not removed produces problem growing areas.

Color Plate 3 (Figure 5.2.) *Improper distribution can create severe application damage.*

Color Plate 4 (Figure 5.8.) *Calibration accuracy is critical.*

Color Plate 5 (Figure 5.9.) Armyworm damage can appear as drought stress.

Color Plate 6 (Figure 5.10.) Weed species invasions can be agricultural versus turf-type in grow-in.

Color Plate 7 (**Figure 5.12.**) *Hand roguing plays a key role in establishment purity.*

Color Plate 8 (**Figure 5.13.**) *Application uniformity is critical.*

Color Plate 9 (Figure 6.1.) Watering to the point of runoff is the goal.

Color Plate 10 (Figure 7.7.) Tires have served as catch basin protection in some applications, making cleanup and maintenance easier.

Color Plate 11 (Figure 8.1.) *Washout repair usually starts in the 'final stage' of grow-in.*

Color Plate 12 (Figure 8.5.) *Slicing and fertilization of weak areas promotes the most rapid turf*

Color Plate 13 (Figure 8.7.) Grow-in layer is experienced on all high-sand greens.

Color Plate 14 (Figure 8.8.) Grow-in layer is evident by touch before sight.

*Color Plate 15 (**Figure 8.13.**) The steel drag mat can be too aggressive on immature turf.*

*Color Plate 16 (**Figure 9.1.**) Wet springs correction are a part of first year maintenance.*

tribution and excellent seed-soil contact because of the cultipacker roller. Rolling in seed after application is *always* preferred to ensure the best seed-soil contact and prevent seed movement.

Smaller seeds such as common bermudagrass, bluegrass, or bentgrasses are not as easily seeded to proper depths. The cultipacker type seeders are set up with a small seed hopper and can distribute hulled, common bermudagrass quite effectively. The bentgrasses, however, require a separate technique of drop-seeding ideally followed by brushing the seed into the upper soil surface. In fairways this can be done with a light drag mat. On greens this is best accomplished with a fan-type leaf rake with no down pressure applied. The idea is to simply provide maximum seed-soil contact without putting the very small seeds too deep into the ground. A .125 inch depth is a target versus the old standard .25 inch.

Remember that when seeds germinate they have a given reserve of food source (endosperm) inside the seed coat that feeds the shoot (coleoptile) until it emerges from the ground and begins photosynthesis. If the seed is planted too deeply, the initial shoot runs out of food reserve before it is able to break the surface and begin photosynthesis. The smaller the seed, of course, the smaller the amount of endosperm and consequently the more sensitive to planting depth.

Bentgrass putting greens, as previously mentioned, are best drop-seeded but then are usually tracked in with the dimpled tire mechanical rake. More moisture is held in the bottom of the dimples for better germination and then this "waffle" board effect is washed out through irrigation and mowing. The waffle board appearance seems extreme at first but does wash out during the establishment phase as seen in Figure 4.9. The most efficient and uniform germination seeding technique is drop-seeding, raking with a fan rake and then rolling with a light walk-behind roller, usually a tennis court type roller only one-third to one-half full of water. Regardless of technique, whether seeds are raked in or tracked in, the final preparation of rolling is by far the most beneficial for germination. Rolling maximizes seed-soil contact and in most cases will improve the rapidness of germination by 10% to 20%. Rolling does make a difference and this demonstrates why

Figure 4.9. *Dimple tracking of seed on putting greens.*

cultipacker seeding is so successful on a large-scale basis. Rolling firms the surface while eliminating the "fluff" on the surface. This, in turn, improves seed/soil contact and reduces the tendency to dry out on the soil surface (Figures 4.10 and 4.11).

Also important is the use of a greens grade organic based fertilizer as a bulk material for bentgrass seeding if the increased volume, seeding technique is chosen. Milorganite® has been the standard for years, but the organics actually are better because particle size and bulk density matches bent seed closer than Milorganite®. Corn meal has also been used for quite some time as a carrier. As a rule, people usually mix seed and carrier 1:3, respectively.

Another technique for seeding bentgrass on putting greens or tees is hydroseeding. Hydroseeding offers very good seeding flexibility, but when hydroseeding bentgrasses, the process cannot be done in one operation—seed, fertilizer, and mulch. The seeds are so small and lightweight that oftentimes they are held up in the mulch cover itself. Consequently a high percentage of the seeds germinate above ground and then quickly die. If hydroseeding/mulching is to be used with bentgrasses, it is best done in a split

Figure 4.10. *Fan raking for best seed/soil contact.*

Figure 4.11. *Rolling after seeding and raking provides greatest germination.*

115

operation of applying the seed first either through hydroseeding or a drop-seeder and then hydromulching on top of the seed once they are rolled in. This gives the most uniform and complete germination, and yet takes advantage of the hydroseeding technique.

One very specific seeding situation which requires its own specialized set of considerations is the seeding of bentgrass fairways. When bentgrass fairways are part of the design, the contours of the fairway cut must be established and well flagged by the architect prior to any seeding operation. Once this is done, the seeding is a two-step process between the bentgrass in the fairways and the other cool season grass being used in the immediate and deep roughs.

The bentgrass fairway perimeters must be established as closely as possible according to design. If fairway contouring is changed after establishment, it is very difficult to move the bentgrass fairway interface in either direction due to the stark difference in turfgrass species. Spots of bentgrass, for example, in bluegrass, ryegrass, or fescue roughs is a noxious weed problem and very unsightly for aesthetics and uniformity. Therefore bentgrass fairway seeding requires very specialized preplanning to ensure the most successful definitions. Figure 4.12 is an example of a bentgrass fairway that has been flagged prior to seeding to identify proper contouring.

Zoysiagrass

The preferred method of zoysiagrass establishment is from sod but this is very expensive. Establishment with plugs has been researched, but plug installation is very time-consuming. Current research shows some interesting results with zoysiagrass establishment from sprigs.

Studies in Maryland show that a zoysiagrass rate is normally about 6 bushels per 1,000 sq. ft. and excellent results were obtained with establishment from sprigs when a proper preemerge herbicide selection was made. Research indicated that oxadizon-

Figure 4.12. Contours must be established before seeding bentgrass fairways.

treated zoysiagrass sprig areas had much greater coverage than nontreated areas and it was the best performer in the preemerge herbicide screen.

Seeded zoysiagrass is another market being closely examined for the potential of establishing zoysiagrass fairways. Current research in Illinois has shown that about 90% cover can be obtained in 15 weeks from Korean common zoysiagrass seed when it is chemically scarified and lightly treated. Weed control in these seeded areas was then maintained with Siduron on a pre/post-emerge basis.

Seeded Bermudagrass

The seeded bermudagrass selections have been significantly improved over the last few years. The better seeded varieties are excellent in areas such as: color, texture, density, and tolerance to closer mowing. Seeded common bermudagrasses today can offer a much less expensive alternative to establishment over hybrid bermudagrass sprigging in many situations where sprigging is not feasible or affordable.

A downside to seeded common bermudagrass establishment is a slightly longer period of time before a playable turf is established versus a turf established with hybrid bermudagrass sprigs. Weed control is also more difficult in the initial establishment phases. Very good results have been achieved in this area with Siduron on a pre/post-emergent basis.

The turf industry's leading seed suppliers have excellent guidelines for seeding rates and special concerns as well as varietal performance which can also be referenced in the NTEP publications. Seeded zoysiagrasses and bermudagrasses both offer excellent performance as plant breeders have improved varietal selections.

Aesthetic Grass Selections

Roughs

Oftentimes, different mixtures are selected for roughs because the architectural idea is to have a textural difference. A common example of this would be the choice of a tall fescue/bluegrass mixture versus a bluegrass/ryegrass mixture. These mixture differences may also be considered for color, such as the use of sheep fescue in deep roughs for a more bluish green appearance, or the use of centipedegrass accents in the southern market for its lighter green color in contrast to bermudagrass.

One area of varietal selection in the southern market that must be carefully weighed is the use of seeded common bermudagrasses in the roughs and sprigged hybrid bermudagrasses on tees and fairways. This is an ideal situation for hybrid bermudagrass contamination from the seeded common bermudagrass during establishment. This can easily happen and therefore these dangers must be carefully weighed when considering a common type in the immediate rough. If seeded common bermudagrasses were used in deep roughs with the intermediate roughs in the same hybrid as the fairways, this would offer a buffer for protection of contamination. Hopefully, the seeding dates would come well before the sprigging date so that a rouging and spot treating with glyphosate in the fairways is done prior to sprigging to further help control the contamination problem between these two grasses.

Naturalized Areas

Naturalized or native grass sites, or meadow grass areas will sometimes consist of a wide variety of different grass species based on adaptability. For example, buffalograss is used in a mixture in arid regions, and this can be with or in place of the fine fescues. Agronomists well versed in native or meadow grass and wildflower blends should be consulted for the proper varietal selection ratios and seeding rates for naturalized blends in a particular golf course environment.

Oftentimes during renovation, native grasses or naturalized areas are established on the golf course in keeping with the updated design theme. The same may be true in adjacent, undisturbed areas of new golf course construction using native or naturalized grasses and/or wildflowers (Figure 4.13).

When establishing such areas, a no-tillage method is recommended whenever possible because this technique discourages germination of weed seeds present. Drill-type seeding is the preferred method, and seeding should be strictly by recommended rates. Excessive seeding with native grasses or wildflowers will result in a poor stand.

Weed control is very important in native or naturalized areas as well. Some of the best means of weed control for weeds that cannot effectively be sprayed postemergent involves mowing, select rouging by hand on a regular basis, and wick applications of a nonselective herbicide. Mowing at the proper heights is most effective on naturalized and native grasses such as lovegrasses or fine fescue mixtures.

Figure 4.13. *Naturalized areas on the course promote wildlife and bird habitat.*

Open burning at the proper time of the year continues to be effectively used for naturalized and native areas. The feasibility of this will be dictated according to whether burn permits can be obtained and whether management is set up to properly contain open field burning. Open field burning promotes improved weed control, thatch/organic matter control, and a healthier environment for native grasses. It is quite often the only weed control needed. Native grasses are more prolific and dense, thus improving aesthetics and wildlife habitat. Research shows open field burning and possibly one mowing per year at 8–12 inches, depending on species mix, is the most productive management for native habitat promotion.

When determining the proper weed control selection, research must be done in naturalized and native areas. Environmental consultants/native or naturalized species specialists can be helpful. For example, milkweed may be a very obnoxious plant, but it plays a critical role in the reproductive cycle of monarch butterflies. This is just one of many examples where research should be consulted when extensive naturalized or native areas are desired. Environmental impact as well as aesthetic qualities are evaluated.

Input from experienced environmental/wildlife consultants is important when planning naturalized corridors for wildlife enhancement around the course. The emphasis on ecotones, the transition areas between various habitat types is a primary design feature. For example, naturally forested areas may be adjacent riparian areas to create a more complete corridor for wildlife habitat. Riparian areas on the golf course would include the edges of creeks or lakes, while the transition from these shoreline areas into heavier forested areas would be classified as an ecotone.

Furthermore, when evaluating forested areas for wildlife habitat corridors, all components of a forested area must be considered. Canopies, understory growth, and even secondary forested growth all play an important role in the overall complete health of a wildlife environment. Therefore, before complete thinning, raising canopies or removing significant understory plants for improved air movement is done, these areas should be evaluated for the entire wildlife environment.

CHAPTER FIVE

FERTILIZATION/CONTROL PRODUCT MANAGEMENT

Spreader/Sprayer Calibration

Accurate spreader and sprayer calibration is critically important during the establishment phase. Turfgrass in the early stages of establishment is much more susceptible to damage from normally routine applications of fertilizers or control products, and the slightest misapplication through overlap or poor material calibration can be very damaging. Too great an overlap or application rate can severely burn tender, new turfgrass plants during establishment. Key calibration concerns must be well identified prior to control product application.

Before detailing control product/fertility management, a primer on equipment calibration is in order. Careful spreader calibration is a common practice in turf management. Sprayers are usually recalibrated periodically to maintain accuracy in control product application. However, spreader calibration is something not discussed or appreciated nearly to the degree it deserves. There has been a tremendous amount of setback in turfgrass establishment through spreader miscalibration in overapplying a fertilizer material or, in some cases, underapplying a material. There are several details that should be outlined when discussing proper fertilizer applications with respect to application technique and spreader calibration.

The normal approach for spreader calibration is to apply a material over a given area, then weigh the difference of the material left in the hopper or collect the material and weigh it to determine the application rate per 1,000 sq. ft. The spreader setting is adjusted accordingly on a trial and error basis until the right application amount is obtained. Then the spreader is calibrated and ready for application. Or is it? If the proper amount per given area of a spreader is calibrated and the desired rate achieved, the spreader is then only 50% calibrated!

Uniformity of distribution of material delivery from the spreader is a part of calibration that is poorly understood. The adjustable third hole on a rotary push-type spreader is there for a purpose—to make sure that the fertilizer is spread uniformly across the effective spread width of the pattern. Some fertilizers such as a finer particle or heavier materials require partial closing of the third hole to balance the distribution (Figure 5.1).

A calibration kit can determine the rate of product per area, and is also used to determine the proper uniformity of distribution across the effective spread width. This kit can be used to calibrate a push-behind rotary spreader, a large tractor mount rotary or pendulum-type spreader. The application boxes are spaced accordingly, with the center box being directly under the spreader itself. Each box is then collected and compared with the center vile to develop a pattern. Figure 5.2, Plate 3 is an example of damage from a fertilizer spreader that applied material too heavily on one side.

As the applicator passed back and forth across the green, the spreader applied the material too heavily to the right side in the direction of travel. When the applicator went across the green and turned to his left to parallel the last pass, the light application side was thrown to the light application side so no damage occurred—it was not an overapplication. However, when the applicator

Figure 5.1. Pattern distribution—an overlooked part of calibration.

turned to his right and then paralleled his last spreader pass with the heavy side to the heavy side, a significant overapplication of fertilizer occurred and thus damaged the turf. If pattern uniformity calibration had been done, this discrepancy would have been seen and damage avoided. This is why it is absolutely critical to not only calibrate a spreader for its output rate, but also pattern distribution. Until both have been checked, a spreader is <u>not</u> calibrated.

Another point to consider with rotary spreader calibration is calibrating with a measurement gauge to where the width of the hole opening in the bottom of the spreader is measured versus the pointer arm of the rate control knob. This arm can be bent or may vary from spreader to spreader, but if a spreader is calibrated by use of a calibration gauge where the hole opening is measured, then all spreaders of that same model will apply the same material at virtually the same rate, because the actual physical opening has been calibrated. Regular recalibration is important with all materials to ensure the maintenance of the application rate, but this is a more accurate calibration that allows uniformity in product application between spreaders of the same model.

125

The effective spread width, by definition, is the distance from the center of the spreader out to one side where the application rate is reduced by 50%. This distance is then multiplied by two to get the effective spread width from left to right. The calibration kit will indicate if this effective spread width is wider on one side than the other because of improper distribution. The effective spread width must be known so proper pattern spacing between parallel passes can be determined. Effective spread width allows for overlap of the total pattern application to prevent under- or overfertilization.

When a spreader misapplication occurs, the superintendent may have to determine which product caused the damage. This is especially true during grow-in when such frequent applications are made to putting surfaces. Spreader/sprayer pattern is an excellent way to identify a problem when it arises. Each time a material is applied, the direction of travel with the spreader or sprayer should be changed, the same as mower direction is changed with each mowing. Using the clock face as a reference for application as with mowing, the direction of application should be noted in the log book with the material being applied. This record keeping allows a problem product to be quickly identified by the direction of pattern streaking and thus a problem material and factors that may have caused the damage can be more quickly identified.

Although this does not allow for correction of the problem because the damage has already been done, it does prevent future problems by pointing out an equipment malfunction or identifying a product with burn potential so use of that product can be evaluated.

Manufacturers today have provided the industry with excellent greens grade fertilizers which are very fine in particle size and very uniform in nature. These products are excellent because they quickly work into the tight, dense turf of putting greens, are virtually undetectable by the eye, and have very little mower pickup after syringing. Many of the materials are also dark in color because of iron and manganese coatings or because of being natural organics. Therefore, their distribution by the eye is almost impossible to see. Consequently, it is mandatory that application spacing be done

Figure 5.3. *Uniform distribution is most important during establishment.*

with some type of marking system so proper parallel spacing between spreader passes can be achieved. The most common way to determine spacing is the use of the dew pattern early in the morning. Using a set of parallel flags on each side of the green spaced according to the effective spread width, as demonstrated in Figure 5.3, is another example. This allows the applicator to push the spreader from flag to flag across the green and not depend on seeing the pattern of the material on the surface and guess at spreader spacing.

There have been instances where spreader passes were made too wide apart and then in effect underfertilized the turf directly between parallel spreader passes. Consequently, *Anthracnose* developed in the underfertilized strips, since *Anthracnose* is a secondary infection disease prevalent when turfgrasses are already under some type of other stress. Had the fertilizer spacing been done with a marking system this could have been avoided. Even though grass was not lost, turf in these areas was set back due to a lack of proper nutrition and reducing its aggressive establishment.

Figure 5.4. *Correct application technique is critical.*

[One side note is appropriate here when discussing spray applications on new greens. In Appendix 8 a walk boom for the spray equipment is required for preparing for spray applications on newly seeded greens. The predominant reason for using a walk boom is that a self-contained, riding unit will probably be too heavy for the greens in the initial stages.] (Figure 5.4).

Fertilizer Programming

Preplant fertilizer and soil amendment applications based on soil tests have already been detailed at length. Additionally, the application of slow release nitrogen at the time of planting has also been outlined to offer needed N after germination. This will help offset the wet conditions of early establishment which can be a limiting factor in granular applications, and the risk of rutting.

Once the initial establishment phase is complete, mowing and fertilization must begin. Initial fertilization raises a lot of questions for grow-in managers who have not undergone the grow-in process before. The biggest questions that arise at this point are: (1) type of materials, (2) timing, (3) soluble vs. slow release, (4) nitrogen vs. complete fertilizer, (5) quantity and frequency, and (6) when to start.

The basic concept of successful grow-in fertility is ideally on a "spoon-fed" program. A balance of granular and fertigation (discussed later) works well together because a steady and aggressive growth rate can be established *and maintained* without the peaks and valleys that can easily creep into an all-granular fertility program. This problem is tolerable to some degree in a maintenance fertility program, but in an establishment situation it can greatly inhibit turfgrass maturity, development, and density.

A combination of soluble and slowly soluble/slow-release nitrogen sources provides the best overall nitrogen fertility management. However, with spoon-feeding, soluble materials provide the overall best response applied lighter and more frequently. Some of the more successful fertility programs have included the use of products with about a 30% to 35% slow-release component early in grow-in to provide excellent initial response and then to maintain some long-term feeding. However, less expensive, soluble products can be used most effectively when applied on a more frequent basis and/or used in conjunction with fertigation. Based on a spoon-feeding theory, this allows for the most aggressive turfgrass growth without the peaks and valleys that can occur when fertility applications are too far apart.

The fertilizer budget during grow-in can be significant, and requires careful shopping by the grow-in superintendent to find the best value of fertilizer per unit of nutrient. Several other factors can greatly improve the fertilizer budget during grow-in: the two main factors being bulk purchase or spreading of granular products, and the use of standard analysis by a manufacturer instead of custom blending. Many superintendents are unnecessarily hung-up on custom blending when often a custom blend is unnecessary. Just because a 25:5:10 with $1\frac{1}{2}$% magnesium is recommended does not mean that a 22:6:9 with 1% magnesium won't do equally as good a job. One fertilizer manufacturer said that they had at one time 110 different 16:4:8 labels registered because of customer insistence on custom blends with variations that would be completely undetectable.

Custom blends are important to supply specific needs based on soil tests or other factors, but the advantages of custom blending in many respects have been way oversold and overemphasized by turf managers. Grow-in fertility does not require fertilizer materials to be overly specific or detailed in their analysis because of the bulk of materials being applied and the frequency. Specific soil needs should have been applied at preplant and the initial grow-in phase is more a need to supply basic nutrients for establishment. Once the grow-in period ends, follow-up soil tests are taken and specific fertilizer programs and analyses are established. Insisting on a custom blended grow-in program is totally cost-ineffective and unnecessary in most cases. Utilizing standard blends offers cost savings due to supplier availability, and a more cost-effective fertilizer budget is the most important focus for grow-in fertility supply. Whether a fertilizer is a 19-3-12 or a 18-4-14 is insignificant during grow-in. The key factor is supplying specific needs that soil tests identify.

All turf managers have seen the impact phosphorous made during initial establishment and its importance at the time of seeding as mentioned in Chapter 4. Of course, starter fertilizers are predominantly phosphorous based, but many grow-in managers do not comprehend the true value of phosphorous and what levels are cost-effective as a preplant. Appendix 13 illustrates why higher amounts of phosphorous are necessary at initial establishment. It is nothing more than a simple matter of exposure of root mass within the root zone.

Oftentimes turf managers will add little or no phosphorous during off-site mixing of fertilizers into root zones in new construction because of a concern over increased *Poa annua* encroachment. Annual bluegrass is favored over creeping bentgrass when phosphorous levels are high in the soil because of better seed germination. Sand based greens have very little available phosphorous naturally occurring and phosphorous is essential for establishment.

Proper phosphorous applications will not induce more *Poa annua* but will enhance turfgrass establishment.

Established turfgrasses have a deep and fibrous root system which is able to draw water and nutrients from significant volumes of the soil profile. This is the very purpose of an aggressive root system—to use the largest soil volume possible for available water and nutrients. An emerging turfgrass seedling has virtually no root exposure in the initial phases; thus, high concentrations of phosphorous in the soil solution are necessary for a seedling to get the needed amounts of phosphorous for assistance in establishment. Soil scientists recommend a concentration of about 15–20 ppm for normal turfgrass maintenance, but suggest a level of 50–70 ppm for initial establishment to offset this lack of exposure.

Fundamentals of Turfgrass Management (4) provides an excellent explanation of the role of phosphorous during establishment as well as other nutrients and the roles they play through maturity and maintenance. This would provide an excellent basis to explain the critical nature of different nutrients and their ratios at different phases of turf development and assist in weighing a particular soil condition, compared with preplant recommendations, to provide balanced soil fertility.

Soil Microbes

Many turf managers believe that soils that are predominantly sand or soils that have received a high amount of pesticides have virtually no microorganism population. This has been shown to be completely false through research. Microorganisms are ubiquitous in the environment and can repopulate even a sterilized root zone media rather quickly. It is true that microbial *activity* is lower in high-sand root zones because of the nature of the sand and the higher inputs of fertilizer, water, traffic, and chemicals, but adding more microorganisms back to the soil *alone* is not the answer.

Soil microbial activity comes predominantly in the form of bacteria, actinomycetes, and fungi. Bacteria are usually the most abundant microorganisms in the soil but other forms play an important role in soil activity. Bacteria activity in the soil have oftentimes been associated with a problem, because bacteria has played an important role in the development of black layer and has been associated with a C-5 decline of Toronto creeping bentgrass which is the only known bacterial disease of turfgrass. However, bacteria are important in the decaying of the more resistant materials found in organic matter and thatch such as lignin and cellulose. They are very important in organic matter breakdown into humic materials.

Mycorrhizal fungi are another form of soil microbiology which develop a symbiotic relationship with plant roots. In mycorrhizal relationships, the fungus benefits from carbon that the plant provides, while the plants benefit from increased phosphorous and water availability that the fungus produces. Research has confirmed that mycorrhizae relationships exist on bentgrasses and bluegrasses.

Research has also shown that under certain circumstances turfgrass establishment was enhanced by inoculation with mycorrhizal fungi, and these differences were apparent within three weeks after seeding. Mycorrhizal fungi have been shown to have naturally colonized new green root zones without being added as an inoculum.

Microbial activity accounts for approximately 70% to 80% of a soil's metabolism (27). As discussed above, however, the turfgrass environment, because of various management needs, is not an environment most conducive to aggressive microbial activity. Environmental conditions that favor good turfgrass growth such as aeration, reduced chemical applications, reduced traffic, and humate/carbohydrate soil management also favor active microbial populations.

In research analysis, humates appear to affect roots more than aboveground plant parts. They also have favorable effects on nutrient uptake and content for each of the major inorganic fertilizer elements including nitrogen, phosphorous, and potassium. These results indicate that using humic substances on soils low in organic matter (sand) gives the highest potential for response. Furthermore, applying humic materials in conjunction with nutrients during grow-in on high-sand putting greens offers even greater BMPs for fertilizer and irrigation management.

Research has also shown that humates can prevent chlorosis in immature maize plants by increasing the uptake of magnesium and iron. Similarly, creeping bentgrass receiving humic substances during environmental stress periods has exhibited better color and fewer dollar spot infestations than nontreated areas, regardless of nitrogen fertilization regimes.

It is recommended that organic based nitrogen fertilizers be an integral part of the overall fertility program during the establishment and maintenance phases of turfgrass management. Organic fertilizers can provide an excellent source for microbial development, iron, and other micronutrients during the grow-in period. Additional micronutrient applications may be required according to soil tests, but periodic organic nitrogen-based fertilizer products can provide a major portion. Evaluating a fertility/soil management program with microbial health in mind will produce improved turf vigor and environmental efficiency.

Fertigation

Fertigation is a management tool which offers a tremendous amount of flexibility and versatility to aid in nutrient application. Fertigation is nothing more than simply applying liquid fertilizers through the irrigation system. The concept has been used for many years in the agricultural market, especially the center pivot irrigation arena. It has also been used in the turf industry for over 20 years, but its popularity has remained mostly in Florida and in the southwest. However, over the last few years fertigation has expanded into all regions of the country in the turfgrass industry. While many misconceptions of fertigation exist, the most common include: (1) it can't be used with an old irrigation system, (2) it will corrode heads, (3) it can't be used in high rainfall areas of the country, (4) it can't be used in the north, and (5) it is impossible to know where the fertilizer is in the pipes. All of these doubts are

Figure 5.5. *Fertigation allows nutrient application immediately after planting.*

completely unfounded. Fertigation offers a wonderful management tool to effectively and uniformly apply a wide spectrum of soil conditioners or nutrients on the golf course at a very low cost in labor and materials.

Today's irrigation technology has maximized the benefits and flexibility of fertigation to an even greater degree. Water management programming through computer controlled irrigation allows for very uniform applications of water, and thus, very uniform applications of nutrients, if fertigation is properly calibrated. Years ago fertigation was difficult to apply uniformly to turf because of the inability to conform the fertigation injection to the output of the pump station. However, with variable speed injection pumps and the ability to tie them into the pump station's flow meter, uniform injection applications are absolute today with fertigation systems. Regardless of the pump station flow rate, fertilizer concentration in the irrigation water remains constant (Figure 5.5).

Appendix 14 is a sample worksheet to develop a fertigation use plan. After looking through this worksheet and consulting with the irrigation designer as to a purge and recharge irrigation cycle, it

Figure 5.6. Microinjection can be an excellent nutrient management tool.

is evident that understanding and calibrating a fertigation cycle is really much easier than most realize.

The argument that fertigation is not applicable in the north or in parts of the country with high rainfall is again a mistake. Fertigation has more application than just bulk N, P, and K from huge capacity storage tanks. Fertigation can very effectively apply soil conditioners as well as micronutrients and other such materials as humates, carbohydrates, and wetting agents. This type of microinjection which applies a small volume of materials at small application rates has been very efficient and very cost-effective. For example, a soil conditioner product is applied at 1 gallon per acre for a cost of about $16 per acre. This application can be calibrated and set up to run overnight with about 20 minutes of the grow-in manager's time, and 80 to 100 acres can be applied in a cycle. That is cost-effective in anyone's thinking (Figure 5.6).

Research over the years has shown that in some instances, liquid applications of soluble materials can be as much as 300% more efficient in plant uptake than granular applications at supplying nutritional needs. All superintendents have utilized soluble spoon-feeding with greens, collars, or tees and have seen the significant benefits. Fertigation offers this flexibility, but its greatest asset occurs during grow-in.

Grow-in and initial establishment is most affected by moisture levels in the upper one-fourth to one-half inch of soil to encourage rapid initial germination of seed or initial rooting of sprigs. Allowing the soil to dry out in the first two- to three-week period is significantly detrimental to the overall establishment rate. Overly saturated soil can also be detrimental. Maintaining high soil moisture requires careful management.

Turfgrasses have a fertility need in the first 2–3 week period that is difficult to completely supply as a preplant. Fertigation fills this niche in program management because nutritional applications can be made with fertigation without having to incur the "drying out for granular equipment" juggling act during initial establishment. It is very difficult to balance the drying process to accommodate application equipment, and then get the water turned back on before drying out or desiccation becomes too detrimental.

Fertigation does not completely replace granular fertility by any means, but the best fertility management program combines the two. Basic soil nutrition requirements are best supplied through a granular program and then supplemental needs based on turf response, environmental conditions, amounts of traffic, or other stresses can be supplemented through the fertigation system. For example, a Tifway bermudagrass fairway is commonly grown in during the 8- to 12-week grow-in period after sprigging with about 1.25 lb. of nitrogen per 1,000 sq. ft. per week. If fertigation is used in conjunction with granular applications, the establishment of a Tifway bermudagrass area can usually be done with about 20% less fertilizer. Normally, if 1.25 lb. N per 1,000 sq. ft. per week is applied on a straight granular program, this is reduced to about .8 lb. N per 1,000 sq. ft. per week granular and another .25 lb. N per 1,000 sq. ft. per week is provided through fertigation. This is a small amount of savings on a weekly basis, but the cumulative effect is significant because the grow-in time will be reduced if proper management and proper fertilizer material selection are utilized. This is due to a more uniform and balanced growth rate through the "spoon-feeding" concept.

Grow-in managers have had similar improved results with applying soil conditioners, micronutrients, etc. with the fertigation system and basic fertility as a granular program. Either method offers significant labor savings and ease of nutritional management. Fertigation should be considered in any construction project and its assets carefully weighed.

Fertilizer movement or nonpoint source pollution is a primary concern during grow-in, requiring careful management of irrigation and fertility.

There are several steps that can reduce the potential for off-site movement of fertilizer materials. These would include:

- use nitrogen sources that are partially slow release
- when using readily available materials, apply at lower rates
- load spreaders on hard surfaces, for ease of cleanup
- if applying soluble fertilizers, load them on turf areas
- when washing spreaders or sprayers, do it on turf areas to prevent runoff
- use buffer strips around environmentally sensitive sites, such as lakes and wetlands
- utilize fertigation for spoon-feeding

The fertility guidelines in Appendix 15 are simply that—*guidelines*. These fertility regimes have proved most successful over the years in providing balanced growth and development based on analysis ratio variation and scheduling to meet the entire plants' needs, not just excessive nitrogen for foliage development. Furthermore, the time sequence for these nutrient programs have proved most successful to coordinate the spoon-feeding concept already developed. Fertility management during grow-in goes hand in hand with water management to keep the most active turfgrass plant growth rate possible. Oftentimes, the fertility philosophy for grow-in is simplified far too much, in that high nitrogen, medium potassium, and some phosphorous, frequently applied is all that is needed after the starter fertilizer at planting.

Hopefully, a better understanding was obtained on soil fertility to quantify availability and then apply the outlined fertility regimes according to *true* need. Tissue testing (discussed shortly) will balance the grow-in manager's understanding and respect of a complete and well-planned fertility program based on actual need. These fertility guidelines are simply compilations of many grow-in programs that have been most effective over the years in various soil conditions and various turfgrass types. All nutrients, not just high N, are carefully rotated. Refer to the bermudagrass grow-in recommendations in Appendix 15. Notice that higher than normal applications of manganese and magnesium as well as sulfate-based nitrogen sources are the foundations for these grow-in programs. Over the years, in sandier soils of southern and coastal climates, these types of base fertility supplements have proved more effective in establishment and health than a similar analysis with only a urea-based nitrogen source and without some of the specialized supplements.

Similarly, a urea formaldehyde nitrogen source is not usually recommended for a fall grow-in because of its sensitivity to temperature and poor availability from low soil microbial activity due to low temperature. Many grow-in managers have problems when utilizing fall and early spring nitrogen applications with a urea formaldehyde base because their response to these materials has been poor. As soil temperatures warm in the spring, an excessive flush of growth prior to the summer heat can be experienced. Consequently poor turf health results. Going into the summer heat in a weakened state increases disease activity, and the extra foliage growth is done at the expense of, instead of in conjunction with, root growth.

These are examples of specific needs of grow-in fertility programs that must be well researched when selecting the proper materials and at the same time finding the materials that are most cost-effective. As mentioned earlier, the slow-release nitrogen portion of a grow-in regime is normally recommended to be about

30% to 35% of the nitrogen balance. However, in the early stages of fertility when greater frequency is preferred, N can be supplied totally by soluble sources if application rates and frequencies are maintained to match N source performance. High soluble N application rates are unacceptable due to the leaching and runoff potential. A single application of soluble N should not exceed 1 to 1.25 lb. N/1,000 sq. ft. and should immediately be syringed to reduce the runoff potential. As the maturity phase progresses, the slow-release component can be increased as the need for grow-in aggressiveness is reduced.

Appendix 15 fertility guidelines may seem excessive at first glance. However, when the program rates are calculated, the total pounds of nutrients are reasonable. These guidelines stress light, frequent applications which produce the most aggressive and consistent growth.

Fertility applications on a four- to five-day frequency is slightly more labor intensive but the goal is to elevate the balanced growth rate and then maintain that rate for rapid establishment and maturity. Increased establishment rate and maturity far outweigh the added labor needed for extra applications initially.

Avoiding an exaggerated growth curve of peaks and valleys between fertility applications must be avoided for best results. Avoiding extremely high fertility rates in any one application reduces the chance of leaching losses or nonsource point pollution from runoff.

The first year fertility maintenance normally consists of about 15% to 20% more fertilizer than routine maintenance programs, just as the grow-in phase is usually 40% to 60% more than maintenance programs.

Soil testing is first done with a complete preplant soil analysis. All soil type areas of the golf course, as well as the green and tee mix is thoroughly checked. The next soil fertility analysis should be when grow-in is 75% to 80% complete. Be sure that soil samples are taken a minimum of 14 days after the last fertilization. Soil samples should be a thorough representation taken about 3 to 4 inches deep. The same soil series samples on the golf course that were taken at preplant should be checked again. A representation of three to five tees, six to nine greens, and three to five fairways is sufficient for this second set of soil samples.

Generally speaking, soil fertility levels will be at their lowest in the late fall and at their best in early spring. This should be kept in mind when evaluating soil samples for nutritional trends. A cool season grow-in should probably have the initial soil samples taken in early to midsummer, with the second follow-up set taken in late fall at the end of the growing season. Warm season grass golf courses, however, will probably have the preplant samples taken in early spring with the follow-up set taken in midfall. The second set of follow-up samples will allow the grow-in manager to evaluate the need for follow-up soil amendment applications (i.e., lime, sulfur, or gypsum), so that by the next active growing season the soil fertility balance can be as complete as possible. This is why the first year's budget after grow-in should have additional fertility dollars available, which to some may seem like overfertilization. However, when considering that fertility level needs will be higher with the specialty fertilizer applications outlined in Chapter 9 and the potential need for additional soil amendments, it is evident that these increased fertility budgets are almost always necessary in the first year's maintenance.

Tissue Testing

A new management tool for monitoring fertility effectiveness during grow-in as well as maintenance is the adaptability of tissue testing to the turf industry. In the last few years, tissue testing has significantly increased in its value to the turf industry because of tissue testing analyses and truer nutrient content values that have validity. There was much discrepancy over the past years as to proper values in turfgrasses and what levels were toxic or deficient. These values have been well quantified and documented with the onset of wet lab improvements for turf as well as a reputable Near Infrared Reflectance Spectroscopy (NIRS) tissue analysis. Both of these programs provide tissue testing as a very useable and efficient tool for monitoring fertilization effectiveness.

Applying fertilization according to soil tests and on experience or projected needs of turf is effective, but to also monitor tissue analysis on a regular basis to determine if nutrients are finding their way into the soil solution and thus into the plant is especially helpful. This is the most effective, best management practice during grow-in and maintenance phases. Appendix 16 has a suggested tissue nutrient value range for some of the key turfgrasses based on current data and from tissue samples accumulated across the country under various management regimes. Regular tissue analysis allows for true custom blending of soluble nutrients to be applied to meet specific needs. For example, if improved color is targeted at some point in grow-in and tissue analysis shows a low magnesium level, then magnesium sulfate might be a better choice for a color response to supply a needed nutrient versus the continued use of iron sulfate for color. Nitrogen levels in tissue can be closely monitored so a very aggressive growth rate can be maintained without excessive nitrogen applications. For example, if nitrogen tissue levels and color are adequate, then maybe the next scheduled nitrogen application would be at .80 lb. of N per 1,000 sq. ft. instead of 1 lb. of N per 1,000 sq. ft.

Tissue testing does not replace soil testing, but complements it to provide a reliable balance sheet for nutrients applied and present in the soil versus plant absorbed. This is an excellent management tool for maximizing the environmental stewardship of turf management fertility programs without sacrificing rapid establishment or maximum turf performance.

Fertility Scheduling

When deciding on the first fertilization after seeding, if fertigation is not being utilized, the grow-in manager should anticipate the estimated sequence needed based on the material applied at seeding. Refer back to the application of a slow-release nitrogen with a starter fertilizer to provide nitrogen during this initial germination or establishment phase. It is generally recommended that when turf reaches about one inch in height, it will begin to show signs of nitrogen deficiencies (4).

Besides the first checkup of soil amendments after grow-in, special needs for fertilization of the golf course must be considered. Appendix 15 has an example of a first year fertility program for bentgrass greens to be compared with the grow-in fertility and a normal maintenance fertility regime. Under most conditions, the grow-in fertility budget on high-sand greens will be about 85% to 100% more fertilizer than a regular maintenance fertility budget. This is gradually scaled down over two years after seeding. However, the first year's fertility schedule on high-sand greens will still be about 25% to 30% higher than a normal maintenance fertility regime because of the increased infiltration in the new and compacting root zone media and because of the increased need for maturity and density establishment. Remember that close water management is the key to prevent leaching losses in high-sand greens.

A common mistake made by many grow-in managers during the later stages of grow-in and the first year's maintenance is to overemphasize nitrogen. Although carefully monitored and balanced levels of nitrogen are critical at any stage of turfgrass maintenance they are especially critical at this stage of development. Nitrogen, however, is not the only important factor in developing an effective fertility program, although it does have the single greatest factor in its effects on turfgrass growth. The single most important factor is a balance of all nutrient needs so nitrogen is kept at aggressive levels yet proportioned with other fertility needs.

All other areas of the course will require special nutritional needs in the first year maintenance, generally having slightly higher requirements for fertility to complete the maturity process and the formation of a dense sod. Usually, in the soil environment versus high-sand environment, the first year's fertility budget will be 10% to 20% higher than a normal maintenance fertility regime based on soil types and locale. Again, these are based on updated soil analyses and the degree of development and maturity established during the initial grow-in phase.

Figure 5.7. *Certain areas require very specialized fertilization.*

There are some areas on the golf course that require very special attention during the first year's maintenance to establish as strong a turf as needed. Weak areas that lack density and maturity such as bunker faces, green and tee slopes, and tees capped with sand require the greatest amount of "baby-sitting" during the first year (Figure 5.7).

Specialized fertilizer applications are needed on these areas in addition to normal maintenance broadcast applications, because of additional leaching, runoff, and possibly exposure stresses due to grade. Usually one to two select applications of fertilizer on slope areas or in the sand-capped tees provide a greatly improved maturity to the turf and development of the root system. This is not to say that one or two extra applications of N alone are needed, but possibly a balanced fertility based on soil tests will provide extra fertility to the more difficult growth areas for maximum maturity and establishment during the first year of maintenance.

The first year's fertility can also utilize fertigation to a great extent to provide some of the additionally needed nutrient supplements required above normal maintenance fertility. This may be an application of iron, micronutrients, or possibly soil conditioners applied simply and cost-effectively with fertigation to all areas of the course, which eliminates costly, special blend granulars to meet these needs. Fertigation, as discussed previously offers an excellent balance with a granular program for providing a more complete fertility schedule for turfgrass.

One other aspect of additional fertilizations in thin or weak areas, depressions (birdbaths), and in settled trenches being releveled with topdressing is the timing of fertility, discussed in Chapter 9. Preplanning a soluble application of nitrogen prior to a drastic cultural program for increasing establishment or leveling an area with topdressing is critical. An application of fertilizer should be made four to six days prior to the operation being carried out so that turfgrass growth is at a maximum, and recovery time minimum. Ammonium sulfate ($AmSO_4$) is an excellent nitrogen source for this application. It produces an excellent growth response and is also very cost-effective. Ammonium nitrate ($AmNO_3$) or urea have been the soluble nitrogen standards for years and $AmSO_4$ has been very aggressively used in the southern or high-sand regions of the country. However, grow-in superintendents in many areas of the country are finding that $AmSO_4$ is a wonderful soluble nitrogen source that produces a better turfgrass response in most circumstances than $AmNO_3$ or urea alone. The sulfur content has been shown to be a needed nutrient in establishment. Nitrate N can also potentially leach more readily from high-sand root zones than ammonium N.

Specialized applications for weak areas are better scheduled at about .75 to .80 lb. of N per 1,000 sq. ft. per application and supplemented on a multiple application basis versus applying 1.25 to 1.5 lb. of N per 1,000 sq. ft. per application. The extra labor involved with a lighter and more frequent schedule is well worth the improved response and subsequent density and maturity of the turf in these still establishing areas.

Control Products

When developing a control product regime during grow-in, the key factors to remember are:

1. read all labels carefully to determine product recommendations and restrictions for newly establishing grasses
2. *triple-check* rates for newly established grasses to prevent burning
3. *triple-check* precautions for newly established grasses; i.e., watering in and tank mixing
4. the calibration of the spray equipment must be absolute every time an application is made.

Figure 5.8, Plate 4 is an example where young establishing bentgrass was badly burned with a fungicide when in normal maintenance no burn would have occurred. An inadequate amount of water as a carrier was applied per 1,000 sq. ft. and the application made during the warmer time of the day in early summer. It is recommended during grow-in and establishment that 3 gallons of water per 1,000 sq. ft. should always be used for control product applications. Research shows that a greater amount of water provides better control and coverage and greatly reduces the tendency for foliar burn. In the case of the herbicide burn, the application used was the labeled rate for established grass instead of immature turf and thus significant setback occurred.

Insect Control

There is no greater opportunity to establish and maintain an IPM program for insect infestation than during grow-in. Since grow-in is the most difficult time to make a control product application because of the very wet conditions and the difficulty of application, insecticides during grow-in should be handled on a curative basis, monitored solely by effective scouting programs.

The exception to this scouting program is testing for nematodes in the greens mix prior to planting. Oftentimes, nematodes can be a problem in the sand greens or tee mix, especially if the source of sand is a river sand. Nematode assays should be screened with the initial testing of the mix for nutrients, and if nematode levels are found then treatment should be planned.

The nematode treatment can be managed from two aspects: if the greens are to be sterilized with a soil sterilant, it will destroy the weed seed, disease spores, and nematodes. However, if the greens are not sterilized, then a topical application of a nematicide should be put down and raked into the upper portions of the soil, followed by irrigation—all according to label specifications.

Scouting is the key for control product management during grow-in establishment because grub and fall armyworm damage can be most severe on new turf. IPM calls for routine scouting to monitor grub infestations, especially in spring as damage is less apparent then because of good soil moisture. The threshold level for Japanese beetle grubs, for example, is usually 10 per sq. ft., but establishing turf lacks the maturity to tolerate root damage.

Fall armyworm activity can literally occur overnight during grow-in and must be closely scouted daily. Young worms are so small they can virtually go unnoticed until significant damage has occurred. The damage appears as a "lack of water" at first glance. Fall armyworms rarely kill grass, but *establishing* grass is a much greater concern because of an underdeveloped root system (Figure 5.9, Plate 5).

The recommended threshold level for fall armyworms is 4 to 5 per sq. ft. in turf, and soap flush can help identify true numbers. A late afternoon treatment is best to match up with feeding habits. If possible, mow the day of treatment, but do not mow for 2 days after treatment.

Under most environments fall armyworms are the most damaging insect pest during grow-in. These worm species thrive on the environment created by an aggressive grow-in program and can quickly become lethal if their presence is not detected for a two- to three-day period. Proper scouting can easily detect the beginning of any activity of surface feeding worms, and spot applications made for IPM programming.

Some areas allow chemigation through the fertigation system, a possibility for large area insecticide application. This is based on an insecticide product being available that is labeled for chemigation application. Local codes and restrictions must be thoroughly researched.

All other insect problems are managed by daily, organized scouting of the entire golf course. Do not overlook the importance of insect scouting during grow-in. Establishing turf is an excellent "target" for insect damage.

Weed Control

The control of weeds in immature, establishing turf can be one of the most difficult jobs of the grow-in management process. Warm season grasses, in particular hybrid bermudagrass and zoysiagrass, have oxadiazon available at the time of planting for nearly complete preemerge weed control during establishment. However, seeded turfgrasses do not have this flexibility although siduron has been excellent for controlling weeds during the establishment of most all seeded grasses. This material can be applied prior to, at the time of, or after seeding for effective crabgrass control.

The best strategy for maximum weed control potential is to seed grasses at a time when weed seedling competition will be minimum—establishing warm season grasses in late spring or early summer and establishing cool season grasses in the late summer or early fall. Seeding turfgrasses out of their ideal seeding environment is a common problem in golf course construction/renovation. The most predominant scenario is spring seeding bentgrass greens and collars on southern courses. However, owners need to understand that weed infestations due to establishment out of season is much more prevalent, which is an important area for the grow-in manager to explain. This updates the owner on what to expect in the appearance and in the possible need for additional postemerge herbicide treatments after the turfgrass has matured enough to withstand applications.

Proper weed control during establishment is a recognized BMP which reduces competition from weeds and allows for a denser turfgrass stand development. This, in turn, reduces soil erosion, leaching, and produces a healthier turfgrass stand which requires less control products and less fertilizer for grow-in.

When a herbicide treatment is required for postemergent weed control in immature turf, triple reading of the label is always sound advice. Establishing turfgrasses have a very low stress tolerance to herbicide, fertilizer, drought, or even traffic. Therefore, a slight mistake in application method or rate can cause significant damage to immature plants. Not only is proper weed identification needed to select the best material available, but also triple checking rates and restrictions for application to immature turfgrass is essential. The other limiting factor is the selection of safe and effective postemergent herbicide materials that are labeled for newly seeded turfs. Generally when applying postemergence herbicides, wait until the turfgrass area has been mowed at least three times or according to other label restrictions before making herbicide applications. Some materials have a reduced labeled rate for newly established turf. Iloxon states that it can be used on labeled turfgrasses after four inches of stolon growth has been achieved, while other chemicals have greater restrictions. These restriction examples must be well understood and identified by the grow-in manager before postemergence herbicides are applied.

Bromoxynil is the only broadleaf postemergence herbicide to date that can safely be applied to seedling turf. It has excellent broad spectrum control on many broadleaf weeds and is labeled on many turfgrass species. Again, herbicide selection is very limited in establishing turfgrasses.

The organic arsenicals such as MSMA can be used on young grassy weeds, but application must be delayed six to eight weeks after turf seed germination. This delay minimizes the toxicity to grass seedlings. Consult the label for rates and restrictions. Local pesticide product availability must be checked as well.

Mowing plays a big part of weed control in establishing turf. Proper height and frequency can greatly increase the turf stand's density, which is the best weed control. Vertical mowing is a cultural means of controlling many weeds in immature turf. Setting the vertical mower to cut about 1 inch above the soil surface is a proven means of physically removing many weed species without damaging the establishing turf. Setting the vertical mower units too low can cause crown damage to turfgrasses. Although some weeds, such as nutsedge, do not respond to the vertical mower, crabgrass or many of the broadleafs are removed quite well by this technique.

Many weed pressures during grow-in are not competitive concerns in established turf. These are more "agricultural" in nature versus "turf" related weed problems. Mowing alone removes many of these initial weeds and the vertical mower will generally remove the balance. Oftentimes during grow-in, weed infestations look worse than they are, but upon close scouting it becomes evident that the weed proliferation is generally a low aggressive population. Figure 5.10, Plate 6 is an example of significant grow-in weed infestations that seem overwhelming at first. In this example, over 90% of the weeds were removed by mowing alone. The more agricultural weeds do not tolerate close, frequent mowing because of their physiology. This is especially common on old pasture land or agriculture fields.

A weed species inventory must be well documented to determine weed problems, corrective actions, and when to implement. The only action necessary could be mowing. If postemergence herbicides are needed, weed population identification is the only way to make the correct herbicide selection. Timing and rate are the last part of the decision process. Broadleaf weed control is best done by spot treating whenever possible. Blanket herbicide applications should not be made on establishing turf because of the potential damage and possible material runoff from an unexpected rainstorm and lack of a dense turf cover. This philosophy is a basic IPM practice. Postemergent herbicide applications must be thoroughly evaluated before use on *establishing* turf. Weed species identification, especially agricultural versus turf weeds, is vital.

Figure 5.11. *Mowing removes most weeds during grow-in.*

Excellent references are available for agricultural weed identification (21,24) (Figure 5.11).

Another factor to implement in an effective weed control program is routine scouting to monitor weed infestations and their aggressiveness. This allows the grow-in manager to identify the weed problems present and to determine how actively they are growing and/or spreading. The scouting process also involves hand rouging on greens and collars for weed invasions. This is effective in removing problem weeds such as crabgrass or goosegrass or even small spots of *Poa annua*. A spot of *Poa annua* the size of a quarter can be dug out with a pocket knife and then the turf pulled back together. Weekly rouging greens during establishment, whether they be bentgrass or bermudagrass, is tremendously effective in removing any off-types of turf that show up. Fairway bermudagrass hybrids find their way to putting greens and collars and require continuous rouging. Rouging is also a recommended practice for bentgrass tees. This is excellent training for assistants, spray technicians, or turf students (Figure 5.12, Plate 7).

Preemergent summer annual weed control combats weed encroachment so prevalent the first year after grow-in due to a lack of turf density. Aggressive weed control is an excellent BMP for improved turf density, long-term improved turf health, and reduction in E&S damage. Careful selection of the preemergent herbicide materials is required the first spring so turfgrass development is not reduced.

For example, oxadiazon [on bermudagrass and zoysiagrass] is usually applied at sprigging as a combination product with a starter fertilizer. This material provides excellent control of most grassy weeds from the time of planting, without inhibiting bermudagrass or zoysiagrass development. This same material can be put down the next spring without inhibiting turfgrass density or development. The most common timing in this situation is a late summer sprigging of the warm season grasses with the oxadiazon application made the following spring.

Cool season grasses are more immature and are poorer in density the first spring after a previous fall seeding. Preemergent herbicides should be used to offset the greater summer annual weed invasions, but material selection must be made carefully so turfgrass development and density is not inhibited. The degree of establishment is the other criteria of a later fall seeding. Here, turf density may be so thin that interseeding is required. This decision is necessary before the preemergent application so proper agronomic programs are developed.

One last area of weed control questions the need for fumigation of high-sand green or tee mixtures. As a rule of thumb, fumigation is needed more on river sands than quarried sands because of a river sand's exposure to weed seed.

Methyl bromide has been the standard soil fumigant for many, many years. Surface-applied granulars or liquid materials also provide excellent sterilization, are less toxic than methyl bromide, and are easier to apply. Most do not require sealed tenting as does methyl bromide, but do require incorporation and irrigation to seal the material in the soil. However, tents laid over the soil surface do provide additional effectiveness for some of these materials. Manufacturer specifications must be carefully weighed when selecting a fumigation material in terms of application rates preferred, application method, and ensuring that the label includes control of the targeted weeds, diseases, and nematodes. A germination test after the reentry period is also recommended.

These types of fumigation materials are very temperature sensitive, and the activity period and the wait period for seeding varies greatly based on soil temperature. This has caused problems for many grow-in managers who scheduled a four-day wait period as for methyl bromide, when in fact they needed a 20 to 22 day wait period for a surface-applied material put down in cooler temperatures. This would ruin the seeding schedule for the greens and could have moved the seeding time out of the desired window. Closely check labels and correlate the wait times based on temperature to the projected seeding time. This can eliminate a poorly planned coordination between fumigation and seeding. It should be emphasized that fumigation is best done by a licensed contractor.

Disease Control

Fungicide applications are a combination of curative and preventative programs. The best initial source for fungus protection during grow-in is using treated seed whenever possible. Treated seed is a tremendous advantage for fungus prevention of *Pythium* and damping-off. Research has shown that a turf stand from treated seed is healthier than a nontreated established area. This improved health continued as much as one year after seeding. Beyond the use of fungicide treated seed there must be a thorough, daily disease scouting program. Remember that fungus environments may be at their greatest during grow-in because the grow-in process involves high amounts of water and nitrogen for establishment, and these are two of the biggest factors which encourage disease activity.

Daily scouting disease on susceptible areas is critical. Young, immature grass plants cannot survive disease attacks and early detection and treatment is far more critical now than in the maintenance stage. Some diseases that are only a nuisance to mature turf can be devastating on immature turf. The areas of disease concern will vary according to grass types and environments. In the southern climates, for example, bermudagrass tees, fairways, and roughs are of little concern, but bentgrass greens and collars are especially susceptible to disease outbreaks during late summer or early fall establishment. This is even more a premium with spring establishment of bentgrass in the southern markets.

On the other hand, northern golf courses with 100% cool season grass require regular scouting over the entire golf course. A combination of preventative; i.e., on greens and tees and curative; i.e., on fairways and roughs, must be used according to budget and environmental stewardship.

Most grow-in managers, as mentioned, use a combination of preventatives and curatives to be cost-effective. High-sand greens, collars, and tees require a preventative program the first few months to maximize health, grow-in, and maturity. If a superintendent normally has a completely curative fungicide program, he must consider the greater susceptibility to disease damage during grow-in and heed the necessity for increased protection during the maturity process.

The same conditions exist on cool season fairways and immediate roughs. At least an eight- to ten-week period for preventatives should be planned. This additional cost is usually justified by increased establishment health. Increased establishment is environmentally responsible because of reduced erosion and nonpoint source pollution by runoff from poorly developed turf areas.

Warm season golf courses usually have minimal fungicide needs during grow-in or maintenance. Careful monitoring on a curative basis is needed, however, because of the "ideal" disease development environment created during grow-in.

One last point to discuss about fungicide applications during the initial grow-in is the use of granular materials versus spray materials. Granular fungicide materials generally offer an excellent alternative when spray applications are impossible. Grow-in managers should always plan to have a spray unit available with a walking boom for control product applications. This also provides nutritional needs when spoon-feeding with soluble fertilizers. Walk booms allow applications to be made with the preferred spray while eliminating the weight of a self-contained boom unit. An excellent planning technique for disease management would be to have one application of granular *Pythium* control and one application of broad spectrum granular fungicide always on hand during the grow-in. Then, if for some reason the spray unit cannot be used, a backup plan is always in place. However, spraying fungicides on the establishing turf with a walking boom is the better selection.

The main disease concerns in any grow-in fungicide program are usually brown patch and *Pythium*. This is true in establishment or maintenance. However, secondary diseases such as anthracnose can be a consequent causal agent that attacks an already weakened turf, and this can happen more easily during the establishment phase due to a lack of maturity, as mentioned previously (Figure 5.13, Plate 8).

Growth Regulators

As ironic as it seems, there is a definite place for growth regulators in the grow-in process of a golf course. Research has shown that growth regulators compact turf growth and increase density. They also provide an increased rooting depth and thus are beneficial aspects of turfgrass establishment and maturity.

Growth regulators serve another very beneficial function in balancing the mowing frequencies of adjacent turfgrass areas that normally would have a significantly different mowing frequency and height regime. The best example of this is growing in the putting surface from seed at the same time the collar and green perimeter is being established, or has already been established by sodding. Usually the irrigation system is a common system for the putting surface and the slopes versus the split system discussed in Chapter 2. This creates a very wet environment for the green perimeter during the initial stages of putting green establishment. Mowing can be impossible, yet growth rates are optimal.

This difference in soil wetness is magnified because the green

Figure 5.14. *Putting surface and green complex establishment can occur at different times.*

profile is usually high sand and the initial water requirements are even greater. Using growth regulators on green slopes and bunker perimeters during this initial phase keeps the vertical growth rate of the turfgrass area down and reduces its mowing needs. Without the growth regulators, the grow-in manager is forced to make a difficult "balancing act" decision. He must either: (1) allow the putting surface to dry for a period of time so that the slopes will be dry enough to mow or (2) mow the slopes when the soil is too wet for ideal mowing and cause mechanical damage.

From this you can see how growth regulators would apply to not only green complexes, but also tee slopes, bunker faces, or any difficult-to-mow area. The mowing workload is reduced through frequency of cut, clipping accumulation, and the amount of mowing equipment required. This would vary according to architecture (Figure 5.14). Check label recommendations carefully to determine various growth regulator restrictions for new sod or immature turf and species.

155

Algae Control

Algae development is a concern during grow-in because of the frequent watering during initial stages. Surface algae is prevalent when temperature and humidity conditions are favorable. Summer conditions coupled with the initial grow-in watering can encourage heavy algae buildup.

The beginning of algae development should be spotted quickly due to daily observations of all areas during grow-in. If initial stages of algae are observed, the approach is twofold. First, the position in the grow-in cycle must be evaluated as it may be time to cut back on water amounts, thus slowing, if not eliminating, further algae development. If this is the case, then reducing water quantity and frequency, as well as the application of a contact fungicide for algae control, can be applied and the problem usually solved. If, however, the course is in an early grow-in stage, and heavier amounts of water must be applied, then a more delicate situation exists. The same contact·fungicide for algae reduction should also be used, but carefully reducing irrigation to allow some additional surface drying, without reducing seedling aggressiveness, must be considered. Furthermore, irrigation stations which apply water to shadier or lower areas, where algae development typically first appears, must also be carefully monitored to see if the irrigation can be reduced.

Allowing algae to go unchecked can significantly reduce the turfgrass stand, especially in the seeding of cool season turf areas. As the grow-in process progresses, spiking along with fungicide applications have been most successful. In mature warm season turfgrasses, especially bermudagrass, a 1% solution of Clorox applied at 2 to 3 gallons of solution per 1,000 sq. ft. has provided excellent algae control, *on an experimental basis*, in a very cost-effective and environmentally safe program. However, superintendents should consult their local county extension office for advice to confirm no pesticide laws would be violated.

CHAPTER *SIX*

PROGRAMMED MANAGEMENT AREAS OF ESTABLISHMENT: IRRIGATION, MOWING

Preplant Irrigation Cycling

Irrigation programming and cycling is a primary management function of the grow-in manager. Irrigation which provides proper soil moisture *without* creating runoff is a delicate balance. Basic initial irrigation cycling can be somewhat predetermined before grassing as the system is checked for pressurization and operation. As irrigation heads are being tested on various areas of the course, the grow-in manager can measure the run times on various locations for the duration before runoff begins. This should be evaluated on different soil types within the golf course as well as on soil types on slopes versus level areas. With this timing in mind, the turf manager can better program irrigation cycling because soil absorption rates will be known prior to planting.

Irrigation Scheduling

Determining initial irrigation schedules has many variables that must be well planned before initial run times and frequencies are established for best water management. The two main objectives of grow-in irrigation are to provide enough water to keep the top .25 to .5 inches of soil moist for initial rooting and establishment without applying too much water to create washing and erosion—irrigation just *to the point of runoff*. With this basic primer in mind a grow-in manager can more effectively evaluate the initial establishment phases of irrigation management in terms of weather, frequency, and duration of each cycle (Figure 6.1, Plate 9).

The initial establishment phase for seed or sprigs varies in duration from about 14 to 20 days depending on: (1) temperature, (2) relative humidity, (3) germination rate of the turf species, and (4) whether it is establishment from seed or sprigs. The following is an example of the initial water regime during germination on high sand greens:

6–11 A.M.	*12–6 P.M.*	*12–6 A.M.*
6 min every 2 hr	4 min every hr	6 min

Bermudagrass or zoysiagrass sprigs require immediate water application after planting to provide the best survivability. The viability rate of sprigs is reduced by 50% in some environments when water is held off for only 2–3 hours after planting if weather conditions are severe. Irrigation should be scheduled immediately so desiccation of stolons is as minimal as possible.

Fairway sprigging should be planted by irrigation zones in extreme heat to reduce the wait time for irrigation start. This usually requires planting across the fairway versus lengthwise. Sprigging down the golf hole can incur planting across 4–5 zones simultaneously so the start time for watering may be delayed 2–3 hours on sprigs. Planting by zone significantly reduces the "wait" period.

When "significant" germination has been achieved (about 70% to 80%) or when stolon development and rooting at the node has begun, the initial phase of irrigation should end and the first reduction in water application begun. This is where the most common mistake in water applications during establishment occurs. All grow-in managers know that water reduction is part of the establishment process but there are two distinct ways to reduce the amounts of water and more times than not the wrong one is chosen. The two ways to reduce water are:

1. To reduce the number of application start times, OR
2. To reduce the run time of each cycle.

Both processes theoretically achieve the same result, of reducing water applications, beginning maturity, and encouraging deeper rooting. However, in these initial step-down phases leaving the number of cycles at the same times during the day and reducing the run time of each cycle is the better way of first reducing water application amounts. Why?

Figure 6.2 illustrates a perfect example of what can happen when immature turf is subjected to the wrong approach of water reduction after germination. The straight sand profile magnifies this as well. On these mounds the turf lacked the maturity to withstand the negative effects of drought, and stretching the time between cycles is, in effect, induced drought stress. Irrigation start times must also be coordinated with heat/drought stress times (and evapotranspiration rates) during a 24 hour period. Stress is greater during the heat of the day, consequently requiring a greater number of irrigation cycles than night or early morning. Refer to the suggested schedule guideline above.

Reducing the amount of water applied through the run time of each cycle versus reducing the number of cycles allows the hardening and maturity process to begin and yet maintains the most aggressive growth rate of the immature turf because the physiological effects of "drought stress" are not experienced. After another 10 to 16 days, depending upon the above-mentioned factors, the second step-down in water application can occur. At this time the number of cycles can be reduced and their appropriate run times adjusted

Figure 6.2. Correct irrigation cycle reduction is very important.

accordingly. The establishment difference is amazing when water reduction is properly balanced during initial establishment.

Recent research has shown some interesting aspects of the role of roots when studying the drought tolerance of turfgrasses. Some experiments have determined that as soils begin to dry, shoot growth can become depressed well before any moisture stress is evident. The belief is that some roots near the soil surface experience drought stress and begin to synthesize abscisic acid (ABA). The ABA signal from the roots warns the plant that the soil is becoming dry and it should reduce its rate of water use, thus allocating more energy for root growth. Through this response, plants obviously develop deeper roots, increasing their access to available water. These studies have been observed in other grasses, not specifically turf, but research indicates that this type of phenomenon does occur in turf. This growth-inhibiting hormone is transported to the leaves where it causes the stomas to close, which slows shoot growth. This obviously is why water management, especially during establishment, is so critical to keep proper soil moistures in the upper portions of soil. Establishing turf that is irrigated according to soil/water potentials only subjects the grass to very mild drought stresses between irrigations. This creates a more deeply rooted plant and utilizes less water than turf that is irrigated by a timer and never experiences drought.

Computer technology today offers tremendous irrigation control when combined with irrigation equipment advances. Programming is available to provide constant flow manager control on pump stations for maximum efficiency and pump life. Programming also provides equally impressive control over irrigation cycling and target applications according to area need.

"Cycle and Soak" type programs allow the superintendent to factor in a soil's absorption capacity and prevents overwatering on slopes and poorer infiltration areas. In grow-in this equates to erosion reduction, not just preventing runoff. It also allows multiple turf area needs to be simultaneously met (Figure 6.3).

Figure 6.3. Slope and climate can vary turf irrigation needs tremendously.

Mowing

Mowing sounds like a simple job, and is one not really given its due importance during grow-in. Ironically, proper mowing management in frequencies, cutting height, and condition of the equipment play a vital role in the rate and quality of establishment.

Mowing should begin when the height of the turf has reached about 25% taller on average than the beginning cutting height. The normal philosophy is to not mow more than 33% of the blade off at any one time, but in the first few mowings, this is a little more leaf tissue than should be removed during establishment. For example, if your initial cutting height is one inch, then mowing should begin when the turf is about 1.25 inches and not 1.33 inches. The ideal scenario is to shock the establishing turf as little as possible. Shock might occur through irrigation mismanagement, fertility mismanagement, poor mowing practices, either mowing too low or mowing with a dull mower. This is why even this small difference in grow-in mowing technique can be a factor in improved initial establishment. After the first four to five mowings, the 33% leaf removal rate is not a problem.

The benefits of mowing go well beyond simple cutting to the proper height. Mowing during grow-in on the proper frequency is critically important for smoothing and firming the new, unstable surface from construction. The initial reduction in water volume is usually timed with the beginning of the mowing operation, consequently, beginning soil firmness. The soil must initially be dried out to prevent rutting from the mowing equipment, and at the same time, maintain adequate moisture in the soil, as outlined earlier.

On stoloniferous or rhizomitous grasses, mowing in the proper cutting height range for the turf species is primary for promoting the most rapid lateral growth possible. The goal with grow-in is to try to promote lateral growth as quickly as possible to begin density and maturity. When cutting heights are either too low or high, lateral aggressiveness is severely delayed.

Proper cutting heights for turfgrasses not only have a range that is too low, but also a range that is too high. Maintaining a cutting height that is too high, a common tendency during establishment, greatly inhibits turf development, maturity, and lateral growth development. Young turfgrasses that are mowed too high remain spindly, lack maturity, and have a lesser ability to harden-off. This equates to a lack of density and a severe lack of rhizome, stolon, or tiller development, depending on the growth habit of the turf. The grow-in manager should carefully consult the specifics of grass species and develop a cutting height program that is about 10% to 15% above the upper range of cutting height initially. This height is used for the first few mowings. By the fifth mowing, the cutting height should be down to the upper recommended range for that turfgrass. After two mowings at this height, a gradual reduction should begin to target the mid-range based on circumstances.

The two most common mowing mistakes made by grow-in managers is making the first cut on new turf too high and too late. All turfgrasses have a cutting height range and to be below this cutting height range is detrimental to the grass. Cutting above this ideal height also reduces the overall health of the plant. Remember, initial cutting heights should begin just above the ideal cutting height range and then graduate the cutting height downward into the upper limit of the ideal range as quickly as possible.

The other mistake is waiting too long to start the first mowing. We discussed above the importance of using the 25% rule for leaf blade removal in mowing practices initially versus the 33% rule normally used in maintenance. Superintendents in the initial phases typically wait too long to start the first mowing because it is easy to be fooled by a lack of density in a newly establishing turf area. The first mowing must be made solely on vertical growth and not on density. Figure 6.4 is a typical example of a new green that should have been mowed 3 to 4 days prior to its first mowing. To look at this new putting green it was easy to be fooled into thinking that it was not quite ready. However, the vertical growth of this turf had gone well beyond the 25% leaf removal rule of the targeted first cutting height. Therefore, be care-

Figure 6.4. The first mowing is based on vertical growth—not density.

ful not to allow a lack of density to influence the decision on when to begin mowing. In this case, looks can be deceiving. This must be a 'vertical growth only' decision, with the other factor being surface firmness. Some additional light rolling may be needed prior to mowing, but by the first mowing, surface firmness should be adequate if the guidelines for seedbed preparation and firmness were adhered to.

A common question here is, "Should the goal for mowing this early be the lower range of the cutting height?" This is dependent on the lateral aggressiveness nature of the selected turf species. Stoloniferous grasses should approach the lower *optimum* cutting height recommendations as quickly as possible, because they have the greatest ability to rapidly develop a mature turf. A predominately rhizomitous species responds most aggressively to a lower, mid-range cutting height. Tillering-type grasses should be kept in the middle of the cutting height ranges as their maturity process is longer because of less aggressive growth characteristics.

When evaluating cutting heights during establishment, there

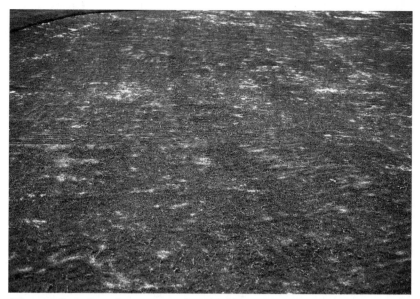

Figure 6.5. Light scalping is part of establishment mowing.

are some specific guidelines to monitor when growing-in a creeping-type turf. After six to eight mowings, the height should be reduced to stay within the recommended cutting height ranges, but ironically, a properly lowered height can actually produce some scalped spots in the establishing turf. Figure 6.5 shows a perfect example of a sprigged Tifway bermudagrass tee that experienced some scalping spots. This indicates how the mowing operation can serve to smooth and firm, because high spots left during the planting process were the only areas being scalped. The smoothing and promotion of lateral growth is thus being enhanced by the mowing operation. As indicated, this scalping is not severe, but only to remove slightly high spots present in newly establishing turf areas. Only creeping-type grasses can tolerate this aggressive approach, but slightly scalping the high spots in rhizomitous and tillering-type grasses, such as ryegrass fairways, can also be a leveling and smoothing tool.

As a general rule, beginning cutting heights as indicated below by grass species/area are:

165

Bentgrass greens/tees	.375 to .50″
Bentgrass fairways	.625″
Bluegrass fairways	1.50″
Ryegrass, zoysiagrass fairways	1.25″
Dwarf bermudagrass greens	.375″
Bermudagrass tees/fairways	.625 to .75″

Mower maintenance is critical during establishment. Reels must be properly back-lapped and spin-ground during the entire grow-in phase. A quality cut of immature turf is more difficult to maintain because of its softness and lush growth, thus it is less susceptible to a quality clip. Because of this factor, maintaining proper mower adjustment is crucial, even though the environment is not the most conducive to reel maintenance. Regular, careful checking of reels must be done to ensure a proper clip. As reels come out of adjustment from the rough ground or soil contact with the bedknife, they "crimp" versus clip the young, tender blades. This increases the shock to the grass and makes it more susceptible to disease. A clean clip must be closely monitored by the grow-in manager, assistants, and mechanics, and adjustments on grow-in equipment—the reel/bedknife contact in particular—must be checked and adjusted regularly. Once per day is not enough.

Frequently checking reel/bedknife adjustment is not usually well-documented or outlined in the grow-in management process, but is a job that can reap as many benefits for quick turf establishment as most any monitoring operation can (Figure 6.6).

To ensure the best cut, mow when the turf is dry. A late morning start is good, with noon being even better. Problems with soil and clipping accumulations on the rollers are reduced, and cutting quality is greatly enhanced. Mowing dry turf during grow-in is a key management program. Scheduling flows smoothly when greens, collars, and tees are mowed in the morning followed by fairways and roughs starting about noon.

Figure 6.6. Flail or rotary mowing may be required to get weeds under control.

Equipment

The initial grow-in phase is very difficult for equipment from soil exposure and the rough surfaces. This is an excellent time to purchase used, walking reel mowers for greens, collars, and, if appropriate, tees. This saves a tremendous amount of money versus purchasing new equipment and has benefited many grow-in managers. Used reels have been purchased for new triplex tractors for grow-in mowing, then rebuilt and used as reels to mow after topdressing or core aerification to protect the newer reels. The used, walk-behind mowers for greens and collars have then been rebuilt and used as back-ups, or mowers used following aerification and topdressing to protect the maintenance fleet. Recommendations for grow-in equipment are outlined in Appendix 8.

Fairway and rough mowing have several different approaches for the best establishment, based largely on grass type. In southern markets where bermudagrass or zoysiagrass is being established, the entire area of fairway, intermediate rough, and deep rough, providing the same grass is in all these areas, should be grown-in as

167

fairway as much as practically possible. This reduces the number of mowers or the different setup for mowers, and is much more productive in the mowing operation. The biggest results are in the firming and smoothing that mowing at fairway height will produce.

Inexperienced grow-in managers often try to establish fairway contouring too early. Raising the rough cutting heights too early in the grow-in reduces turf development and maturity. Too, it creates a situation that requires scalping when maturing the turf at a time when scalping is most detrimental to turf health. Mowing as much as possible at fairway height means that after fairway contours are set by the architect, the roughs can simply be allowed to grow up instead of being scalped down. This philosophy is very different than trying to develop the fairway/rough step-cut too early in the grow-in process. It also eliminates the strain for grow-in mowing equipment to be set at different heights (Figure 6.7).

Bentgrass fairways and bluegrass, ryegrass, or fescue roughs present a different philosophy in mowing equipment selection. Again, triplex reel mowers or the lighter weight five-plex units are recommended in fairways. The simpler belt-drive units have been

Figure 6.7. Mowing most areas as fairway promotes greater maturity and simplifies mowing.

some of the best and most dependable service equipment in the grow-in. Simplicity and durability is of utmost importance in an environment that is so difficult on equipment. Belt driven mowers have performed remarkably well compared with more delicate or hydraulic-driven equipment. Equipment beyond this belt-driven simplicity has not performed significantly better in the grow-in environment when maintenance versus quality of cut is measured.

A combination of pull-behind hydraulic five-plex and large deck rotary mowers has provided the best results for the grow-in of cool season roughs. Oftentimes, a combination of pull-behind hydraulic reel units is being used in the intermediate rough area with large deck rotary mowers used on slopes or deep roughs. Rotary mowers provide an excellent cut on all cool season turfgrasses during establishment but do require a sharp blade at all times, as for reel mowers. Higher RPMs and slower ground speeds also add to cutting quality.

If a predominance of fine fescues is being established in deep roughs for naturalization or very low maintenance areas, they should be cut at the upper recommended cutting height ranges during the establishment phase. As discussed, all turfgrasses have a recommended cutting height range, and for the best establishment, maturity, and rooting development of turfgrasses they should be maintained in this cutting height range until mature. The tendency in naturalized fine fescue areas is to seed and never mow them. Lack of mowing reduces the development and maturity of the grasses. If fine fescue blends or mixes are being established on steep banks, this of course is a limiting factor for mowing during establishment. However, if fine fescues are being established in deep rough areas that allow mowing, the quality of the stand during establishment can be greatly enhanced. These deep roughs may be mowed less frequently than immediate roughs, but mowing does improve establishment and is very beneficial in weed control. An established cutting height of about 4 to 5 inches is recommended in this situation.

Raising cutting heights is a normal agronomic practice on warm season grasses in the late summer to encourage improved carbohydrate storage in the root system for winter dormancy. This

169

is especially important in the first winter after grow-in because of turfgrass immaturity after the first growing season.

Normally, hybrid bermudagrass fairways mowed at .5″ are raised to .75″ in the early fall. Common bermudagrass fairways mowed at about .75″ are normally raised to 1.25″. Warm season grasses should be raised an additional 10% to 15% in cutting height the first winter after grow-in. This extra height provides a greater degree of winter hardiness for warm season grasses as they begin hardening for winter. Proper potassium applications in the early fall will maximize winter dormancy survival.

Not only is raising cutting heights in the fall beneficial to warm season grasses, but it can be beneficial to cool season grasses as well. Cool season grasses in the spring after the previous late summer/fall seeding are immature in most climates and need careful attention at this time. Slightly raising the cutting height (about 10%) the first spring after grow-in allows much greater root development, faster maturation, and better preparation for the upcoming first summer. Even this small increase makes a significant impact on root development and improved maturity.

Row Planting

When converting a common bermudagrass fairway to hybrid bermudagrass by row planting, some specific management practices must be implemented to maximize success. The first practice is spraying the fairways with glyphosate about one week before rowplanting. This stunts the common bermudagrass so it does not compete with the hybrid bermudagrass sprigs until well into the hybrid bermudagrass establishment phase.

Conversion by row planting and cutting height is often not discussed. Common bermudagrass in fairways performs much better at .625″ to .75″. Tifway bermudagrass provides a better playing surface at .5″. A true .5″ cutting height is too low for common bermudagrass unless fairways are exceptionally smooth. However, to mow fairways at .5″ after hybrid bermudagrass sprigs are *estab-*

lished from row planting will allow for a much more aggressive takeover of the hybrid bermudagrass. Continuing to mow the fairways at .625″ to .75″ after row plant establishment allows the common bermudagrass to compete "head to head" with the Tifway bermudagrass, inhibiting the conversion.

Mowing fairways at .5″ will cause some scalping in the common areas and the first summer after row planting may produce fairways that are not as pleasing as the golfing membership desires. Explain the lower cutting height regime to the membership so the scalped areas are anticipated with the understanding that this is necessary the first summer to ensure hybrid bermudagrass takeover.

CHAPTER SEVEN

EROSION AND SEDIMENT
CONTROL/DAMAGE REPAIR

According to the Soil Conservation Service there are several types of erosion by water, and the various types must be understood before erosion control management can be fully understood. (1) Splash erosion occurs as raindrops break the bonds between soil particles, making particles much more vulnerable to movement by water flow. (2) Sheet erosion begins when surface water carries along particles that were detached by raindrops. (3) Rill erosion occurs as surface water quickly establishes its own path. As these concentrated flows or rills begin to experience more cumulative and rapid water movement, soil is dislodged and carried away at an accelerated rate. Other defined forms of erosion are listed as concentrated—flow erosion, gully erosion, and mass erosion or slumping. The first three, however are the primary environmental concerns. Keeping soil in place, not just on the property, is the key to successful E&S control.

The late Dr. Bill Daniel of Purdue University stated that on average, sediment yield from construction sites can be 20 times higher compared to adjacent agricultural watersheds. The amount of sediment lost from a construction site is due primarily to such factors as the length of time ground is exposed, the size of the area exposed, and the topography of the site in terms of velocity and water movement (5).

It is estimated that the U.S. loses 3 billion tons of topsoil per year to erosion. Moreover, the sediment load of the Mississippi River is estimated at 300 million tons per year. Generally the bulk of nitrogen lost in surface runoff is in the form of organic nitrogen associated with eroded soil. Over 80% of water quality problems in estuarine and coastal waters is from nonpoint sources. The main water quality problem in most states is sediment in streams, lakes, or estuaries.

Golf course drainage has become a complicated issue due to concerns of *where storm water goes and what's in it*! The old norm was simply to exhaust surface and subsurface drain water into lakes, streams, or swells that end up in bodies of water. Today, environmental awareness has caused golf course drainage to be more sensitive to where runoff water is directed. However, it must also be pointed out that golf course management by the superintendent—i.e., water, fertility, and control product applications, receive maximum scrutiny through Integrated Pest Management (IPM) and Best Management Practices (BMP) programs for responsible environmental management. Areas such as nonpoint source pollution, storm water pollution prevention plan (SWPPP), BMP, and strict erosion and sediment control (E&S) plans must be submitted, approved, and then rigorously enforced throughout construction, grow-in, and maturity (Figure 7.1).

Some parts of the country are subject to wind erosion. Although wind erosion is not as prevalent or at the magnitude of water erosion, some parts of the country can experience significant wind erosion or damage from dust storms during a construction activity. In the 1930s the American Great Plains basically took to the air. In the U.S. alone, almost 70 million acres have serious wind erosion problems to date. Therefore, wind erosion, although not a consideration by most areas of the country, must be researched in the Great Plains or similar areas as to whether it must be managed for golf course construction. The erosion and sediment control

Figure 7.1. Storm water can create significant damage if not controlled.

plan developed for the golf course will have such factors, where appropriate, as the continual wetting of disturbed areas to prevent dust from becoming airborne.

Erosion, whether by wind or water, is no small factor in its environmental impact. Today, $27 billion of the U.S. total annual losses are estimated through reduced soil productivity from wind and water erosion.

If wind erosion is a consideration for a particular area, the grow-in superintendent or the owner rep must make sure that the greens mix is well protected from soil contamination due to wind erosion. This has been done with silt fence protection around the green perimeters and in some cases, greens may even have to be covered with plastic tarps to protect them from exposure. This should be well researched by the superintendent and local civil engineers to try to determine how significant that damage could be. Syringing the green complex in the later stages of construction prior to planning could also be beneficial in keeping enough moisture in the soil to reduce soil movement from wind erosion.

Soft engineering is a term referring to methods used to prevent catastrophic shoreline erosion. These methods create stability, and the construction materials must be selected for each section of a stream channel based on the flow volume and velocity. Naturalization of the bank and the area above the bank finalizes this program (32).

Damage Repair/Maintenance

Erosion and sediment control (E&S) is a critical part of construction and grow-in. It typically is never discussed in the overall scope of work that grow-in entails. Ironically, E&S control management will be one of the single most time-involved duties, supervision, and maintenance, during the first one-half of the grow-in process.

E&S control management should begin during the project construction/grow-in and its installation and maintenance responsibility belongs to the golf course contractor during construction. The responsibility is usually then turned over to the superintendent when grow-in begins. This section will deal with some specifics of the E&S control process, details involved in managing the E&S control plan, and how to read the E&S control plan ensuring that it is properly implemented in the field.

The details and specifications of any erosion and sediment control plan is drawn up in accordance with state or local agency specifications for E&S control management. A copy of the state's E&S control specifications can be obtained through the Natural Resource Conservation Service. Each state has its own set of specifications and these specifications can be further tightened by the county. A copy of the field manual will describe in detail various structure types, their materials and design features, their role in the E&S control program, and proper maintenance of the structures in the field.

Reading the E&S control plans can be overwhelming to superintendents thrust into this management role for the first time. However, the plans are relatively simple to read and very descriptive in nature so that implementation and monitoring in the field does not require an engineering degree. The material on the back inside cover of the book is an excerpt from an erosion and sediment control plan with various structures drawn in and labeled at appropriate places in the field according to storm water movement patterns. Also there is a corresponding legend sheet which is found with most erosion and sediment control plans and adapted from the state erosion and sediment control manual. In this legend, various structures are named according to type (ds1, ds2, or a diagram of a silt fence). Identification of these various structure symbols on the plan is then compared to the key to see the identification of these structures. Construction material composition and proper installation is detailed in the E&S manual as are maintenance details of each—the most important factor for the grow-in manager. The superintendent will be most concerned with routine maintenance requirements during grow-in and specifics for each structure on the project.

Each structure has very stringent specifications based on function in the field, and many specifics such as gravel or riprap size can make a difference in its function. Careful detail must be given to what types of materials are speced and what maintenance activities and schedules are required. Something as seemingly insignificant as using the wrong size gravel can create an ineffective storm water structure and thus dilute the overall effectiveness of the E&S control plan.

One such structure that is poorly understood is the littoral shelf. The littoral shelf plays a significant role in pond management beyond just serving as a safety ledge. It entails the establishment of a vegetative buffer on the shoreline that serves as a filtration, curbing surface runoff of sediment and nutrients into a body of water. The littoral shelf prevents erosion of an exposed bank subject to tidal fluctuations, including wave erosion from wind at low tide or, if a rain occurs when the water level is down and raindrops directly dislodge the bare soil.

A littoral shelf may also be speced on a small pond that is subject to significant drawdowns during periods of heavy irrigation. If this is speced on the erosion and sediment control plan, this structure serves the same as a water basin subjected to tidal fluctuations.

Littoral shelves may also be speced at pond edges to serve only as the vegetative bio buffer for the introduction of surface runoff water.

Once plan details are identified and well understood, the setup of the maintenance process for the plan can be implemented. The main thrust of the E&S control plan is to keep topsoil in its place and reduce as much soil movement as possible. Many people define the E&S plan as keeping soil on your property. However, the Department of Natural Resources (DNR) looks at proper E&S control management as keeping soil in its place instead of just on the property. Preventing the loss of quality topsoil should be the goal of any property manager.

Proper E&S control management and maintenance requires a very proactive role, such as structure integrity and effectiveness monitored after each rain event. Daily inspection of these structures, done as the grow-in manager is inspecting the golf course, will allow for as-needed repairs if damage has occurred from an irrigation problem like a blowout or a stuck head. When considering the actual definition of erosion and sediment control, it is clearly understood why water management through irrigation goes hand in hand with E&S control management. The proper watering regime for grow-in is light, frequent applications to keep the upper portion of soil moist. However, too heavy an application not only wastes water, but also 'produces' soil erosion.

Water velocity and volume creates erosion damage. The goal of an erosion and sediment control plan is to release storm water at a *controlled rate* to reduce damage. A controlled rate means keeping storm water slowed down and spread out over as large an area as possible. Improperly managed irrigation can quickly cause significant erosion problems because of its concentration of water to a given area. Many grow-in managers have created tremendous damage and erosion during grow-in because of poor water management in the initial grow-in stages. Dealing with erosion washes and surface unevenness in the first year of maintenance is discussed in detail in Chapter 9, but it is important to relate this damage to erosion and sediment control management. Additional work created for the first year's maintenance could have been prevented with proper water management during the initial grow-in.

Figure 7.2. *Properly maintained E&S control plan.*

Figure 7.3. *E&S structure storm damage.*

179

Figure 7.4. *Catch basins serve as sediment basins during construction/grow-in.*

Figures 7.2 and 7.3 show typical examples of E&S control structures on a golf hole where washing is most severe. Figure 7.3 shows an improperly managed silt fence structure and the corresponding damage that can be caused from poor maintenance.

The proper maintenance of catch basins is illustrated in Figure 7.4, and improper maintenance during grow-in is illustrated in Figure 7.5. It is evident that if sediment is allowed to accumulate at a catch basin, it can quickly overcome the rock protection of the stand pipe, depositing silt directly into the storm drain system. This not only eliminates catch basin effectiveness, but also deposits the silt in unwanted areas, and/or can plug up a storm drain system. To unclog a silt plug in a storm drain system is a tremendously difficult job. Handling silt clogs is outlined below. Figure 7.6 further details the end result of silt accumulation not handled properly during final seedbed prep/floating. The silt buildups that occur during construction must be regraded and worked back into the surrounding soil. If not redistributed, a silt layer remains which

Figure 7.5. Unmaintained, silt-damaged outlet.

Figure 7.6. Catch basin area requires proper seedbed preparation.

would be most difficult to establish turf on and would be especially poor at draining because of no infiltration to this area.

Figure 7.7, Plate 10 show examples of one erosion and sediment control operation done during grow-in. Oftentimes, old tires have been effectively used to protect catch basins because they allow water to accumulate and allow silt to settle out of the water before the water builds up and spills over the tire into the storm drain system. Removing the tire and then removing the silt deposit is very easy. Sodding around catch basin inlets during the grow-in process is tremendously beneficial in eliminating damage from moving or swirling water from around the catch basin structure. Discuss this method with local E&S monitoring agents to determine if it is acceptable.

Unplugging Drainlines

Siltation from storms can plug drainlines when the ability to dissipate silt exceeds the drainage system or catch basin capacity. When silt 'bridges' in drainage pipes and plugs them, the tendency is to try to unplug the drainpipes from the catch basin end. Actually, the pipe is more effectively unclogged from the lower side as the silt or debris builds up on the upper side of the pipe and the front edge of the clog becomes a wedge driven tighter in the pipe as the buildup grows. Therefore, unplugging from the front has most effectively been done with a high pressure hose. One of the most commonly used methods for silt unplugging utilizes a one-inch hose attached to a ball valve and using a quick coupler valve in the irrigation system. The ball valve is then reduced to a 1/2 inch PVC pipe and the pipe is inserted into the drain system. This creates a very high pressure, high velocity water jet action. As the pipe is fed into the drainline it begins to dislodge the plug by relieving the pressure and water flow can be reestablished, allowing the drainpipe to flush itself. There are water jet heads designed specifically for such needs and a construction contractor is a good source of obtaining one if desired.

Wash repairs are the most common repair needed with erosion and sediment control. The proper handling of a wash repair varies greatly depending primarily on the severity of the wash damage which can depend on factors such as steepness of the grade and the volume of storm water concentrated into that particular area. Some wash repairs can be handled simply with silt fencing placed immediately above the wash to redistribute storm water over a greater area. More severe washes require redistribution of water over a much greater area *above* the wash; checking dams to further slow down water movement through the area; and then sodding or an erosion blanket to further protect the soil.

When evaluating the proper technique for wash repair, it is important to explain to owners and developers why sod is sometimes chosen over filling a wash with soil and reestablishing with seed/sprigs and mulch. Sod is the most expensive way to repair a wash, or so it seems when the price is first evaluated. But if a wash area has to be repaired once, including cleanup of the silt deposit, refilling the wash with soil, compacting, and leveling the fill, then reseeding and remulching—sod just became a cheaper alternative of fixing the erosion damage. Sod may be more expensive initially, but the cost of sod to eliminate another repair of the same wash in terms of labor and materials is a less expensive alternative as well as a more effective erosion protection and the best BMP method. These high potential wash areas can be identified during final construction so they are protected with sod at *initial* planting. This is another example where field mapping is so beneficial.

When selecting proper erosion repair techniques for an area, the grow-in manager must first correct the source of the problem—the storm water moving across the erosion area. Proper evaluation of how to handle this storm water must be carried out prior to wash repair. Figure 7.8 is an example of properly handling the source of the problem with severe washing. However, notice that it shows extensive wash repair well into the grow-in program. Permanently correcting an erosion problem early in the grow-in is not always possible.

Figure 7.8. Educated wash damage assessment can save on labor.

This is a prime example of erosion damage that is almost impossible to repair until the soil is secured around the wash through turfgrass establishment. In this case, there simply was no other place to channel storm water without creating the same degree of damage in adjacent areas. Therefore, the actual wash was left in place and check dams maintained to keep water velocity slow, so water was channeled through this wash without creating <u>further</u> damage. Then, after turf was established around the wash, the storm water could be redistributed into other areas without damage and the wash repaired. Figure 7.9 depicts exactly how erosion and sediment control areas can become very costly when they are repaired three or four times during the grow-in phase due to improper evaluation and technique. Erosion and sediment control is more than just installing silt fence and filling wash ditches with soil. Carefully evaluating the source of the problem and then determining how best to correct it is an essential part of an effective E&S control plan.

Wash repair and smoothing surface roughness from minor erosion damage are addressed in the later stages of grow-in and

Figure 7.9. *Grass sometimes must be established before total E&S damage can be eliminated.*

during the first year's maintenance. Eliminating surface roughness from limited erosion is discussed in detail in Chapter 8. Remember, however, that many erosion repairs can be properly carried out only after significant turfgrass establishment. This does not mean an erosion area is left unchecked until the grass is established, but it does mean that a damaged area may have water channeled through it during grow-in, but with check dams installed to keep the water velocity down and distributed at a pace that reduces further damage. Again, repair comes after the grow-in.

Temporary Grass Covers

Sometimes construction projects require temporary grassing to stabilize soil until the permanent turf cover can be applied. A temporary vegetative cover may be used from about two weeks to ten months. This is most commonly needed when summer/fall construction requires overwintering before the spring/summer seeding the following year.

Certain requirements should be identified when selecting a temporary grass cover. The most obvious is choosing a cover that would not out-compete the permanent turf cover. Another is selecting a temporary cover that will stabilize the soil as needed. The USDA publishes a handbook (Handbook No. 170) available through the U.S. Printing Office which catalogs many species that perform well as temporary vegetative covers.

Annual ryegrass is the most common selection used as a temporary grass or added to a permanent turfgrass blend/mixture to aid in establishment. Care must be taken to add only a small percentage (10% by weight) of annual ryegrass to a permanent turfgrass mixture. This will ensure the "short-term" presence of the ryegrass.

Some projects use intermediate rye as a winter temporary cover to have the stabilization characteristics of annual ryegrass without its tremendous shoot growth. The ryegrass is seeded at light rates—usually about 150–200 lb./acre as a general cover and 225–250 lb./acre on steeper slopes. Other species such as cereal ryes have also performed well in golf course situations and might be explored.

Temporary grass covers are a valuable part of E&S control programs when longer term soil exposure is necessary. The DNR representative should also be an excellent reference for temporary grass cover suggestions.

Mulching

Mulching is an establishment aid with significant value. It not only improves the germination rate by holding moisture, it is a tremendous deterrent to direct erosion. Mulch reduces direct contact of the soil to water droplets from rain or irrigation. This splash ero-

Figure 7.10. Mulching is an effective E&S control program.

sion dislodges soil particles. Mulch covers "intercept" the water droplets, thus reducing <u>initial</u> erosion that would occur on bare soil (Figure 7.10).

Straw mulching should be put down at 3,000 to 4,000 lb. per acre. Loose mulch can be used on a slope with a 5:1 slope or less, but a 5:1 slope with significant wind exposure should have mulch crimped with a coulter disc, oversprayed with a bonding agent, or anchored with netting. The same would be true for steeper slopes such as a 4:1 or greater. Mulches are also best used in low flow swales, as higher flow rate swales will require an additional means of keeping storm water velocity down.

Hydraulically applied mulches are put down at the rate of 1,500 to 3,000 lb. per acre. Some mulches require a tacifier agent when used on slopes of 5:1 or greater, as is generally recommended.

187

CHAPTER *EIGHT*

CULTURAL PROGRAM
ESTABLISHMENT

Though cultural program management is an integral part of regular turfgrass, it is not often considered to be a significant part of the grow-in and maturity process. Although techniques and aggressiveness may differ from normal maintenance, cultural programs play a significant role in turf maturation and the length of time developing quality playing conditions.

In Chapter 5, the theory of setting the vertical mower at a higher height to improve broadleaf weed control is explained. This is just one example where cultural programming and equipment impact grow-in, but with different techniques than normal maintenance. In this chapter we will outline the various aspects of cultural management and the role each plays in the grow-in and maturity of playing areas. Cultural management will be discussed from its initial use during grow-in to the completion of the first year maintenance. First year cultural programs can vary significantly from normal maintenance. The cultural management programs used during grow-in include topdressing, aeration, vertical mowing, brushing, use of the drag mat, grooming, and rolling.

The topdressing process during grow-in provides complete smoothness to playing surfaces prior to course opening. The turf-grass species dictates the volume of topdressing applied, and this is especially true during grow-in. At the later stage of grow-in both heavy and light topdressings are applied according to need and oftentimes this aggressiveness has been dictated by the amount of erosion damage that occurred during initial grow-in. Poor water management or heavy rain can create extensive rippling on tee and green surrounds during the early establishment, requiring significantly more topdressing and rolling in the later stages of grow-in and in the first year's maintenance. Figure 8.1, Plate 11 is an example of this more-than-normal washboarding that occurred during grow-in, and the need for additional topdressing.

The first year's maintenance involves a great deal of work to achieve quality playing surfaces. It is not uncommon for golf courses to extensively topdress greens, collars, tees, approaches, and sometimes landing areas, depending on budget and equipment restraints. Topdressing can be largely used to improve trench settling and birdbaths in tees and fairways with proper timing and techniques. The alternative to topdressing such areas is the lifting of sod, releveling the area, and then replacing the sod. This is expensive, labor-intensive, and much less productive. If topdressing can be used instead of this leveling and smoothing, then imagine the cost efficiency and the speed with which the job can be completed. The specific details of heavy topdressing must be well planned to ensure success and turf recovery through the topdress application.

Applying very heavy topdressing to repair damaged areas such as in Figure 8.2 requires detail timing and orchestration of fertilization to make this heavy topdressing application successful. Simply "dumping" two inches of soil or sand in an area and then expecting the grass to grow back through just won't happen. Figure 8.2 is a prime example of this. Prior to the first year's growing season following initial grow-in, a careful mapping of the active growing season is needed so three to four evenly spaced, heavy topdressings can be applied during the peak of the growing season. These dates are then identified on the maintenance calendar. Ap-

Figure 8.2. Many washouts can be corrected through well-planned topdressing.

proximately four to five days <u>prior</u> to these topdressings, an application of soluble nitrogen fertilizer such as ammonium sulfate should be put down at 1 lb. of N per 1,000 sq. ft. in the areas to be heavily topdressed. This establishes a very active growth rate, so when the topdressing is applied the turfgrass has a much more rapid ability to grow through and firm up these releveled areas.

Topdressing depths should be confined to about .5 to .75 of an inch per application, depending on turf species, followed by firming with lightweight rolling. After rolling, syringing with a hose followed by rolling again should provide maximum firmness and quick recovery on an already aggressively growing turf.

The most common mistake when topdressing in such a manner is twofold. One, too much sand material may be put down at any one time; and secondly, a grow-in manager will often wait until the sand is applied or even a few days <u>after</u> before fertilizing. The key is to apply the nitrogen source prior to topdressing to stimulate an active recovery growth when the topdress application is made. Again, with careful planning prior to the growing season, three and sometimes four of these applications can be made through the

Figure 8.3. *Birdbaths are common in new construction.*

growing season with great success. The labor savings are tremendous because of a significant reduction in the necessity for sod lifting and releveling.

As discussed previously, topdressing birdbaths and trenches settled in fairways and approaches can be done with the saved spoils of the greens mix stockpile that was contaminated during the mixing and loading. This makes excellent use of this material or of a finer sand used on sand capped tees. Either way, use a sand material that has some soil separates in its makeup (usually a river sand) so it is easy to work with and yet will be firm, not droughty, after the turf is established. A sandy loam type soil has also been used quite effectively in similar type topdressings (Figure 8.3).

Slicing is a major cultivation program used during grow-in to smooth turf areas and promote maximum lateral growth. Slicing benefits creeping and rhizomitous turfgrasses and even tillering types, but it is done to a lesser degree on tillering turfgrasses. Figure 8.4 illustrates slicing in a bermudagrass fairway to further promote lateral growth and improve density. Slicing is also quite effec-

Figure 8.4. *Slicing has tremendous benefits during grow-in.*

tive on new bentgrass tees and fairways—more maturity is required, however, than with bermudagrass.

Slicing has much less surface disruption than aerification and helps to smooth and promote a very strong lateral growth of creeping-type grasses. The surface smoothing and tillering also benefits noncreeping grasses, but the frequency of slicing is much less. On bermudagrass or zoysiagrass fairways, for example, slicing usually begins about five to six weeks after planting and is replicated on a seven- to eight-day frequency through the balance of grow-in. Bentgrass fairways can withstand slicing about 11 to 12 weeks after seeding under normal establishment rates, which can be replicated every 14 to 16 days for the balance of the grow-in. The golf course should be sliced in all areas. This practice is beneficial to sodded areas, too, for smoothing and knitting together. It is further needed for breaking surface tension and to promote better air and water movement through the sod layer and into the profile. Core aerifications of newly sodded areas and their schedules will be discussed later in this chapter.

Slicing is often used during the first year's maintenance to en-

Figure 8.6. Rolling new greens without topdressing is excellent for smoothing/ firming.

courage additional lateral growth in weak areas or areas that did not fully establish density during initial grow-in. Figure 8.5, Plate 12 is a perfect example of a thin area that remains after the initial grow-in, and it is being spot sliced for additional lateral growth. Many grow-in managers choose aerification at this stage. If enough density has been developed to pull a core, this is fine. However, slicing provides the same benefits at this stage of development with less surface disruption.

Rolling greens during grow-in is not considered a cultural management program, but is, in effect, because of the different outcome produced than when used in maintenance. Figure 8.6 shows a self-propelled roller being used during new green establishment. Here, the root zone mix needed additional firming and smoothing to provide the needed root zone stability. Irrigation about 20 to 30 minutes prior to rolling provides the soil moisture needed at the surface.

Many grow-in managers hesitate to roll during these early stages because of the damage incurred to young turfgrasses. Rolling with a tennis court-type roller, or even speed rollers does not damage young turfgrasses if not applied in conjunction with topdressing. The damage from rolling only occurs when it is used immediately following or a short time after topdressing. Abrasion occurs when exposing the still immature turfgrass plants to the sand particles, thus bruising the plants. Rolling as illustrated can firm and smooth putting surfaces to aid the rolling operation of the mower, and reduce the tendency for scalping or gouging that can occur from the front roller of the putting green mower.

A self-propelled, walk-behind roller is best because foot damage from pushing a roller on a delicate or unstable surface is eliminated. The PSI weight of this type roller can be adjusted by the amount of water added. In Figure 8.6, the roller is only about one-half full. It had been used on this same green approximately two to three weeks earlier and at that time was only one-third full. The weight can be adjusted to the circumstances to maximize the benefits of rolling without being detrimental. This type of rolling should be encouraged on greens, collars, tees, and approaches. It is also most beneficial on seed/sod interface lines to make an absolutely level transition. If sharp bank shoulders are design features of a golf course at tee boxes or in greens complexes, this type of roller might also be used to soften the sharpness of this shoulder just enough to prevent mower scalping. A two-tier green is a common example.

This is not an attempt to change the design of the contour feature, but sometimes shoulders are left too sharp, which has been especially a problem on two-tier greens where the tier elevation change was not properly floated and blended together. Thus it was too severe to allow mowing without scalping. Rolling on this slope during establishment will help to soften any severity that may be present. When rolling a two-tier green, it is best to roll in the direction of the tier as well as perpendicular to it. If the tier is not rolled in both directions, it creates an unevenness at the bottom, accentuating the mower scalping. Hand watering and then rolling a tier both longitudinally and latitudinally will be much more beneficial in providing absolute contour smoothness from the top of the tier to the bottom and the (skirt) area immediately under the tier.

Sometimes these tiers or ridges in putting green design are too severe to be mowed at modern-day cutting height without scalp-

195

ing. If this scenario applies, owners will have to be made aware that a slightly higher height of cut will be necessary to accommodate this severity. With the newer bentgrass varieties, this may mean a height of .160″ to .165″ instead of a target of .130″.

Speed rollers or lightweight rollers used for improved putting green speed are also effective in firming new putting surfaces both at seeding and after germination. The point of concern with using these riding rollers is the turning at green's edge to change their pattern track across the green. Some superintendents use ¼″ plywood to make this pivot, while others have been able to make the turn in the sodded collar, depending on green complex design.

Consider too if using the riding rollers on a still-predominantly sand based surface would void a warranty because of potential bearing damage. Also test this type surface to see if the rollers will have sufficient traction for moving across the surface without damage.

Aeration is a key cultural management tool for use in the later stages of grow-in and more extensively during the first year of maintenance. Like slicing, aeration has similar functions in promoting lateral growth and smoothness. Compaction relief is not a concern at this stage of golf course maintenance. However, aggressively aerifying during the first year's maintenance of established turf is a significant part of fine-tuning the smoothness of playing surfaces and completely eliminates construction "scars."

Aerifying fairways, intermediate roughs, and slopes, accompanied by a large drag mat to break up the cores as a topdressing is the best means to remove lingering construction scars from grow-in. Small greens-type aerators pulling larger cores such as a ⅝ inch tine are needed in collars, approaches, and tees. However, the larger pull-behind or three-point hitch units with ⅝ to ¾ inch tines should be used in all other areas for maximum benefit. During the first year's maintenance, all areas should be scheduled for three aerifications to promote the greatest surface smoothness and turf maturity possible. Three aerifications in the first year may seem extensive, but again it should be emphasized that compaction relief is not the objective now. Smoothing surfaces, promoting improved density and greater lateral growth, and providing a simple and easier way of topdressing through the dragging in of cores, are all benefits of aerification in the first year of maintenance.

As mentioned, aerification is better than slicing once turf maturity develops. Spot treating weak areas with additional aerifications during the first year of maintenance promotes the most rapid turf cover. As outlined, when discussing heavier topdressings and rapid turf recovery, the same holds true with fertilization timings here in spot treating weak areas with the aggressive aerification program. The thin areas that receive additional aerifications must also be fertilized five to six days *prior* to the aerification to promote the most rapid healing and also to gain the greatest benefits from the aerification. These fertility programs, recommendations, and the theories behind their timings are discussed in detail in Chapter 5.

One aerification use that is nearly always overlooked during grow-in and the first year's maintenance is to remove the *grow-in layer* on putting greens. ALL putting greens experience a grow-in layer, whether composed of high sand or topsoil. This grow-in layer is nothing more than a rapid accumulation of organic material from the excessive growth rate of the turfgrass, promoted from the very nature of grow-in management.

Consider the whole scenario of grow-in. Grow-in is simply an accelerated program to develop a quality turf as rapidly as possible to allow a golf course to open as soon as possible. To develop this rapid maturity, extensive amounts of fertilizer and water are put down to promote growth. Mowing practices must accommodate these rapid growth rates to encourage density. These programs encourage the accumulation of organic matter because of the soil's inability to break organic matter down at a rate equal to accumulation. This is further enhanced by reduced thatch management programs in the initial stages of grow-in—the lack of topdressing, aerification, and vertical mowing. Bentgrass and bermudagrass are significant thatch producers under normal maintenance regimes.

Layering

The commercial industry today has given the turf manager the means of aerifying in almost any conceivable manner. Turf managers can aerify on less than two-inch centers, 1/4 inch to 1 inch tines, go as deep as 9 or 10 inches, and even aerify with water injection. However, layers in high-sand greens remain as persistent today as they did before this new aerification technology. Therein lies the problem—aerification and topdressing alone are not enough to manage layers. Soil management through soil microbiology promotion to break down organic matter with the physical assistance of cultural programs is the only true answer for layer management. This is discussed in detail in Chapter 5.

A grow-in layer will be experienced by every golf course superintendent after the grow-in period. Managing it early is critical to the long-term health of the green. Figure 8.7, Plate 13 shows evidence of a grow-in layer very early in the life of this new putting green. Figure 8.8, Plate 14 is an example of a grow-in layer of high organic content that is detected by the feel, but undetected by sight. This is why using a profiler such as seen in these examples, versus a smaller .5 to .75 inch soil probe, is more effective for evaluating profile health and putting green performance. This type of layering and root development is impossible to effectively monitor with a small soil probe, which is excellent for checking moisture content or pulling soil samples for analysis. However, for true profile evaluation a profiling-type tool should be utilized by every superintendent (Figure 8.9).

This grow-in layer is very evident by touch, but is visually undetectable until it begins to darken in color. In the first season after grow-in, layer management through aerification, topdressing, and the rigorous use of humates and carbohydrates are all important and should be foundation programs for proper soil management of putting greens. As discussed earlier, aerification alone is not enough to manage this layering problem but it does break the surface tension this layer can create, improves air and water movement through the soil profile, and makes a healthier environment

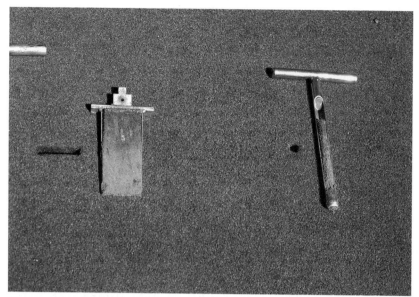

Figure 8.9. The profiler shows soil condition much more effectively.

for microbial activity. However, microbial activity will not thrive in a high-sand root zone without assistance from an improved environment conducive to microbial activity—one which provides the proper food source as part of the fertility program.

Many superintendents today are urged to apply bacteria through the sprayer to improve the "sterile" conditions of high-sand greens. Microbial activity exists in high-sand greens as it does in any soil environment throughout the world. Even soil sterilization is short-term because the microbial population is so ubiquitous. While adding additional microbial populations can be beneficial, it does not solve the source of the problem. A proper growing environment must be established and maintained in the high-sand profile before increased microbial activity can be experienced and the benefits reaped.

High-sand putting greens are exposed to a huge array of factors detrimental to active microbial activity. High traffic, high amounts of water, high fertility, and high control product applications are all factors necessary to meet today's putting green standards. However, they are all significantly detrimental to microbial

199

activity and adding more bacteria, just as aerification alone, is not the answer. Regularly adding a food source will create a better environment for increased microbial populations and activity.

Humate materials and carbohydrates are a great food source and create an environment in any high-sand green that will produce greater microbial activity whether more outside bacteria is being supplemented or not. Agricultural research conducted in the '40s and '50s discovered the benefits of humates and carbohydrates in soil management. The goal of these products is to provide a food source to build up microbial activity.

As microbial activity increases, organic matter breakdown efficiency increases. This reduces layering and improves fertilizer efficiency because of an improved breakdown cycle. In effect, the superintendent needs less fertilizer because of a greater efficiency of the fertilizer material applied. The benefits are almost immeasurable. Putting microbial activity and basic soil nutrition with cation balance as the foundation of all fertility programs yields reduced fertilizer requirements, healthier turfgrasses, and consequently, turfgrasses that will withstand greater stress conditions whether from pests or environmental factors.

Layering is often associated with sodding. One misconception in our industry today is that washed or soilless sod does not have a layer to worry about after establishment. This is, unfortunately, not true because although the layer from introduction of soil may not be present, the layer of pure organic matter from the sod being root-bound is present and can be a greater problem than the grow-in layer from seed or sprigs. Figure 8.10 is a sod layer from soilless sod put on a root zone mix that met USGA specifications. However, the following growing season after sodding, it was not aerified because there "was no compaction." Consequently, the layers in the high organic sod base held tremendous amounts of water at the surface and created a surface tension that in effect created a false perched water table. As a result, the turf had tremendous localized dry spots in some areas and wet wilt in others because of its inability to uniformly manage the water in the root zone and in the pure organic sod layer. Aerification and the introduction of proper soil conditioners would have all but eliminated these terribly inconsis-

Figure 8.10. *Soilless sod has an organic layer to manage.*

tencies. The putting greens would have experienced a rapid organic reduction through natural breakdown, reestablishing the desired uniformity in the profile.

Initial aerifications at the end of grow-in and during the first year's maintenance on high-sand greens seems ridiculous to many. Remember that aerification during this phase of development is not done for compaction, but for smoothing, additional lateral growth promotion, and for grow-in layer removal in high-sand putting greens. Until this management program is appreciated and used in the turf industry, layer management will still fall short when aerification and topdressing alone are depended upon. The soil microbiological management of layers was discussed in more detail in Chapter 5. It must, however, be related back to this misunderstood aerification program for grow-in.

Aerification should be scheduled at the beginning of the growing season following establishment. For example, if greens are seeded in late August with bentgrass, then aerification should be scheduled the following spring as soon as weather allows. Ideally, two aerifications are scheduled that following spring using .25 to .375 inch tines. The goal is not to punch large holes, but instead to punch as many holes as possible to break the surface tension of the grow-in layer, allowing for better water management. When the surface tension of the mat layer is broken by physical disruption, uniformity of air and water movement through the profile is reestablished.

Vertical mowing is not often associated with grow-in and establishment. Used in the proper perspective, however, it offers another excellent tool to promote establishment as rapidly and as uniformly as possible. As outlined in Chapter 5, vertical mowing can be of effective weed control assistance in the grow-in phase, but it also has some very directed establishment and maturity promotion in late grow-in.

In Chapter 6 we discussed the advantages of proper mowing techniques to aid in smoothness. On creeping-type grasses, this smoothing effect can be further enhanced by vertical mowing on collars, tees, and approaches. After significant establishment and density is developed, vertical mowing is used to promote smoothness, but in a different way than in maintenance. This difference is basically twofold: (1) vertical mowing blades should not be as closely spaced as they are for thatch removal in a maintenance operation, and (2) the vertical mowing blades are set slightly above soil level instead of down to, or slightly below soil level. Vertical mowing during this time is often too aggressive, not because of the operation itself or the depth at which the vertical mower blades were set, but mostly because the vertical mower blades are too close together—thus causing crown damage.

Many turfgrasses such as zoysiagrass or colonial bentgrass respond quite well to vertical mowing when done in a proper regime—with the vertical mowing blades being spaced farther apart. Blade spacing on such turfgrasses should be about 1 inch apart instead of .5 to .625 inches. The same technique is true during the later stages of grow-in and in the first year's maintenance. Thatch control is not the target of vertical mowing during maturity, but surface smoothness is. Therefore, more widely spaced vertical mower blades can aid in surface smoothness without being too harsh on establishing turf.

Another aspect of vertical mowing during establishment is setting the vertical mower blades slightly <u>above</u> the soil surface. A bench setting of about .125 inch above the soil surface provides an effective depth of about .0625 inch above the ground which is ideal for this type of vertical mowing operation. Experiment with this bench setting to achieve the proper effective cutting depth, which can vary based on grass type and density achieved.

Bermudagrass greens require additional fine-tuning through cultural practices at the end of grow-in to ensure total surface smoothness. The sprigging operation versus seeding, and the introduction of plant material on the surface can create a surface that is much rougher than that from seeding.

At the end of bermudagrass green grow-in, light yet aggressive vertical mowings followed by topdressings and brushing will offset this additional disruption in surface smoothness. Light refers to the verticutter depth setting—being only to the soil surface. Aggressive refers to the frequency. The vertical mowing will dissipate the clumps of plant material left at the surface from the distribution of the sprigs. Light vertical mowings, ideally with a blade spacing of about .5″ width initially, can begin about 7 to 8 weeks after sprigging under normal conditions. Again, this vertical mowing has a slightly wider spacing than normal vertical mowing for grooming and is not set as deeply into the soil as is the case for thatch removal or grooming. The main objective is to thin the clumps of plant material still present at the surface and maximize surface smoothness.

This setting depth allows for high spots in the soil surface to be broken up by the vertical mower blades, which will also help remove lateral growing leaves and stolons that are not tightly attached to the ground. This, in effect, grooms the surface by creating a more upright growth habit and at the same time removes lateral leaf tissue that will only add to thatch accumulations. Moreover, this operation smoothes the soil surface without damaging turfgrass health (Figure 8.11).

After vertical mowing is complete, drag or brush the surface

Figure 8.11. *Light vertical mowing promotes density and smoothness.*

to distribute the soil from high spots that were broken loose by the vertical mower blades. At this point, the thatch brought to the surface can be swept up or blown off and removed. Topdressing might be added if necessary, followed by another light brushing.

Care must be taken on more tender establishing grass plants such as the bentgrasses or bluegrasses. Brushing or drag matting topdress into the surface can be most detrimental to still immature grass plants. The species mentioned above will not tolerate the abrasiveness of topdressing sand being worked down to the surface as well as bermudagrass. Therefore, it may be best to space the follow-up topdressing after vertical mowing back two to three days to allow some recovery of the initial damage. As detailed earlier, any time these type of operations are carried out during the establishment phase, a soluble fertilizer application should be made four to five days prior to the operation. This greatly aids recovery from such cultural management programs and can increase the recovery rate as much as an estimated 50%.

Under certain circumstances, the use of topdressing in cooperation with brushing has many advantages over the use of the

Figure 8.12. *Brushing is better for grooming initially.*

steel drag mat (Figure 8.12). Generally speaking, the brush will be less abrasive to the establishing putting green turf than the drag mat, because the still maturing grass will not tolerate the aggressiveness of a drag mat and topdressing as well as established turf. Brushing thus allows earlier grooming of the turf and rapid promotion of density and smoothness. As maturity develops, the focus turns to the steel drag mat for aggressive grooming and the brush is a supplemental "fine-tuning" grooming tool. During the first year's maintenance, however, the steel drag mat can be used as in a normal maintenance operation for aggressive smoothing if applicable.

Drag mat damage is sometimes a concern. Actually, the damage from the drag mat more often occurs when used in conjunction with topdressing than the drag mat itself. The abrasiveness of the sand serves as a grit no different than the back-lapping of a reel, and abrasion to the leaves can occur. Established, healthy turf can withstand and quickly heal from some abrasion, but immature turf can be devastated. The drag mat is more effective as a smoothing operation on putting surfaces versus a brush. However, brushing can be used earlier in the establishment operation because it is less damaging (Figure 8.13, Plate 15).

The aggressiveness of the drag mat when used alone varies by the depth of the drag mat size. Drag mats are available in variable sizes, and many are wide but not very deep. This allows for the beneficial effects of the steel drag mat without overexposure of the turf to the drag mat. For example, a drag mat 5 ft. deep instead of 8 ft. deep is much less abrasive to the turf because much less mat would have passed over it. This is the size of a drag mat needed in the late stages of grow-in or early first year because it is less aggressive. Deciding whether multiple passes are needed with the drag mat strictly depends on the maturity and health of the turf stand. Care is taken to not "circle" multiple times over the same area to drag adjacent parts of the green—e.g., a narrow area on the putting surface that is difficult to cover.

The steel drag mat has excellent cultural benefits in the later phases of grow-in and in the first year's maintenance for all turf areas. All turf managers know the benefits of a steel drag mat moving the topdressing material down around the crown area into the soil surface. The drag mat is also helpful in smoothing greens immediately following aerification and topdressing.

The steel drag mat has also been used to groom greens by pulling it over the surface prior to mowing, without a topdressing application. The mat simply "scuffs" the turfgrass up to encourage a more upright growth, as do other vertical mowing and grooming operations, which in turn allows for a cleaner cut.

One other brushing operation of significant benefit in fine-tuning quality playing surface is the use of a rotary-type brush or street sweeper. The rotary brush is an excellent grooming tool in maintaining fairways, intermediate roughs, and tee tops. It provides a means of light vertical mowing without the thatch cleanup, and is often used to sweep surfaces of debris collected in wind rows. The operation works from the center of the fairway out to the immediate roughs and the debris is distributed by the tractor-mounted blower. The rotary brush is also an excellent tool in the southern overseeding market for preparation of a seed bed in bermudagrass turf.

Use of the rotary broom in the later stages of grow-in or in the first year's maintenance allows for additional surface smoothing and removal of small stones and sticks. The down-pressure of the rotary broom can be adjusted to meet the needs of the situation.

Many superintendents create their own drag mats by using a section of chain-link fence. A chain-link fence serves as an excellent drag mat when properly weighted, and a rather large unit can be dragged over large areas for time efficiency. The design of the chain link serves well to break up and redistribute cores and/or topdressing without being overly aggressive to the turfgrass.

The flexible-tine harrow has also been widely used in the turfgrass market and has served as a drag mat following aerification. It has been used in the first year's maintenance and pulled in the least aggressive position. However, the grow-in manager must be careful to assess the use of this type of drag mat in the early stages because of its much heavier construction, and in some cases it can be too aggressive before complete maturity is established.

Aggressive cultural practices and grooming are not usually considered as major parts of golf course development until the second year. On the contrary, in the first growing season after grow-in, aggressive cultural practices and grooming are most beneficial to establishing the best playing surfaces and turfgrass health possible. For warm season grasses, this would be the following spring after a summer planting. For cool season grasses, this would include the spring following a late summer/fall planting.

Early aggressive aerification, vertical mowing, on a grooming basis, topdressing and brushing produce improved turf health and fine-tune playing surfaces at this stage of development. Waiting another growing season in belief that the turf is too immature to withstand these practices sets the maturity back noticeably. In the first growing season after grow-in, the soil is still relatively soft at the surface and thus fine-tuning smoothness can be done better now than waiting until later. Now is the time to remove construction scars and light erosion reels in fairways and immediate roughs by aerification, thoroughly dragging in the cores and then rolling any bumpy areas.

CHAPTER NINE

FIRST YEAR BUDGET/
SPECIAL NEEDS

First Year Maintenance

The first year maintenance of a new golf course or of a major renovation project is almost never discussed. However, first year maintenance operations are quite different than normal maintenance and require some specific planning, programming, and budgeting to accommodate special needs that must be identified and addressed at this stage.

The first year budget differs from a normal maintenance budget because of the specialty needs to finish the maturity process that a new golf course or extensively renovated golf course experiences. These special needs would include such items as: (1) washout and final E&S control repair; (2) unexpected drainage problems, e.g., wet springs; (3) birdbaths; (4) readjusting mowing contours to best accommodate design and play; (5) additional aerification needs; (6) special fertility needs of many areas of the course such as remaining weak areas, slopes, and bunker faces; (7)

tree removal and replacement from construction attrition, shade/air movement evaluations; (8) fine-tuning of all playing surfaces to maximize density, quality, and surface smoothness.

Appendix 18 has an example of a first year budget by line item. This should be used as a guideline for key areas that are not part of normal maintenance and includes some narrative for each to offer suggestions for presentation to ownership. The first year budget offers surprises to those inexperienced in the grow-in or major renovation of a course. Therefore, advanced planning here is as important as any time in the overall development to prevent an inadequate budget.

Wash repair and the correction of birdbaths was discussed at length in Chapter 7. It is important to reemphasize that these items must be budgeted in the first year maintenance budget in order to correct them. A well-planned, instrumented, heavy topdressing is the best approach to repair most washes, trench settling, and birdbaths. These types of repairs are often more extensive on golf courses that experience significant rain damage during grow-in or on golf courses built by nongolf course contractors. Poor water management during grow-in can be another significant contributor to excess damage that must now be repaired during the first year.

The extent of wash damage repair can easily be evaluated at the end of the first growing season. Trench settling and birdbath repairs should be accurately identified after the grow-in season so proper projections for planning and budgeting can be prepared for the upcoming first year maintenance program. These repairs should not be delayed but corrected in the first year maintenance for best results.

Unexpected drainage problems almost always show up in the first rainy/winter season after grow-in, usually as newly surfaced wet springs in fairways or immediate roughs in construction cuts versus fill areas. Figure 9.1, Plate 16 depicts an unexpected wet spring which surfaced during the winter rainy period after initial grow-in. A golf course with any significant cuts during the original construction will experience a few wet springs that surface. These wet springs are simply underground water seeking the path of least resistance and will break ground on a cut slope now having surface exposure. Subsurface drainage will also break above ground in the first significant rainy period when the water table rises.

Wet springs can usually be corrected without significant expense or problem unless there is no place close by to exit a french drain. There is a specific way to dry up wet springs, but many grow-in managers have attempted correction by using french drains with poor results. Let's look at the proper approach in correcting a newly surfaced wet spring.

Remember, subsurface water breaks the surface because it follows the path of least resistance. To correct this with a french drain, however, the drain needs to be installed on the <u>upper</u> side of the wet spring and not the lower side. To put a french drain on the lower side means that the water will have to break the surface, then flow into the french drain, move down into the pipe, and be removed from the area. The proper method to dry up wet springs is to install the french drain above the wet spring, thus intercepting the water <u>before</u> it breaks the surface.

French drains above wet springs should be installed as deep as possible at the spring so it can intercept the water as deep in the ground as possible. Preferably, french drains should be trenched about three feet deep along the wet spring and then tapered out, depending on grade, to normal depths as the french drain is carried to the exit area. French drains should have gravel installed to about three inches from the surface, and then a medium sand capped over the pea gravel. The sod can then be reinstalled, if necessary, but this sod should be rigorously aerified two to three times during the first growing season. This will eliminate any surface tension the sod layer might create, disrupting the ability of the french drain system to properly function. Sodding over drainage lines can render them significantly less effective and sometimes completely ineffective if the sod creates a seal. Appendix 19 is a french drain profile properly constructed.

Silt berm accumulations on cart path edges are shown in Figure 9.2. Correction is a part of any first year maintenance program and must be immediate to prevent breakage of paths due to the soft ground created by the trapped water at the silt buildup. Most berms can be corrected by removing the soil back to the original

Figure 9.2. *Silt berms must be corrected promptly.*

surface grade. Others will require subsurface drainage because of an unanticipated low area that holds water. Whether at curb breaks as in Figure 9.3 or simply at accumulation sites during establishment, silt berms should be corrected as soon as possible to reestablish surface drainage patterns.

The importance of aerification and topdressing was discussed in Chapter 8. There is a significant need for aggressive aerification, slicing, and topdressing during the first year's maintenance, not from a compaction relief need, but for a finish establishment need. Aerification of all areas of the golf course in conjunction with the use of a drag mat as outlined in Chapter 8 is significantly beneficial in providing the desired surface smoothness. Rigorous aerification on collars, approaches, tees, fairways, and immediate roughs is most important. The same is true with green and tee slopes and bunker faces as extensively as design will allow. Aerification in these areas is not as important for smoothing as it is to eliminate the surface tension of sod layering and to create a more uniform soil profile, thus a better rooting media for the turf. Aggressive first year aerification is essential for courses that totally sod fairways

Figure 9.3. Silt can accumulate at curb breaks to interrupt surface flow.

and/or roughs. A sod layer benefits from aerification regardless of location on the course.

The advantages of slicing weak areas during first year maintenance was outlined in Chapter 8. Slicing the remaining thin areas is significantly beneficial in encouraging the most rapid density. Many grow-in managers choose to go exclusively with open spoon aerification tines in the first year maintenance, and this is perfectly acceptable. However, some prefer to use slicing in these weak areas and do so more frequently due to less disruption to the surface. Both programs have done well, but slicing alone over the entire golf course during the first year's maintenance should not be the standard practice. Most of the turf areas will be completely mature and established during the first year and therefore would now benefit the best from hollow tine aerification.

Completing maturity, improving density and surface smoothness are the main objectives in the first year after grow-in. Consequently the aerification/drag mat combination is most beneficial. The same is true with increased topdressings the first year, again to maximize surface smoothness. This is usually accomplished on greens, collars, approaches, tees, and sometimes landing areas. Rigorous topdressings in these areas the first year should always be done in conjunction with aerification to minimize topdress layering buildup.

On tees and approaches, first topdress and then aerify to bring the cores up on top of the already applied topdressing. The surface is then dragged and the soil cores are broken up and incorporated back into the aerification holes as they are blended with the topdressing sand. This reduces the tendency for topdress layering, and this seemingly backward operation has another benefit. During the first year's maintenance, the surface, after aerification, is more unstable than normal established turf areas. Because of this, it is often impossible to use a topdresser after aerification because of tire rutting. The same is true on putting greens. To accommodate this special concern, topdressing prior to aerification prevents tire rutting. Again, after soil cores dry on top of the topdress-covered surface, a drag mat or brush can very effectively break up the cores, blend them with topdressing, and then work this blend back into the surface and the aerifier holes. Figure 9.4 is an example of the operation being carried out on greens for this reason. This program greatly reduces topdress induced layering.

Specialty fertilization is discussed in detail in Chapter 5. It is important to note that when preparing first year budgets, additional fertilizations of key areas such as bunker faces, slopes, and weak areas should be planned for. Normally in first year maintenance, bunker faces and slopes require an additional 10% to 15% fertility for total establishment. These weak areas are often under-fertilized during grow-in because of the difficulty of application, which is magnified during grow-in. Weak area maintenance, depending on the lack of maturity and the acreage of these weak areas, can be evaluated and budgeted for in the off-season after grow-in.

Another area requiring attention in the early stages of the first year's growing season is readjustments to mowing contours. Ideally, contour mowing patterns are established at the end of the grow-in season, provided the seeding/sprigging window was met and ade-

Figure 9.4. *Topdressing before aerification reduces tracking on new greens/tees.*

quate establishment developed. This is done with the architect so mowing patterns are in keeping with design philosophies.

We discussed in Chapter 6 the differences in the mowing contour establishment of bentgrass fairways. During this first year's growing season, adjustments in bentgrass fairway contouring are made, if needed. Hopefully, the initial contouring and seeding was done with great care so adjustments are not necessary. It is more intricate to move contour lines between bentgrass fairways and bluegrass/ryegrass roughs than simply changing the mowing pattern where the same turf is in the fairways and roughs. This often requires an application of Round-up® followed by reseeding of the proper grass. Bentgrass can be a major weed problem in roughs because of its significant difference in appearance and growth habit from other cool season grasses. Therefore, the architect should be encouraged to adjust contours, if needed, at the very beginning or even slightly before the growing season of the first year's maintenance. If necessary, the nonselective touch-up followed by seeding can be done as early in the growing season as possible to allow for establishment time.

215

If the contour changes made should require some scalping, these contours are marked out with paint and the scalping done in a two-step height reduction procedure. Remember, the turfgrass is still somewhat immature and is not as tolerant of significant stress as mature turf. If, for example, you wish to scalp out a rough area currently at 2 inches and the fairway is at .75 inch, mark out this contouring with paint during the architect's visit, then mow the turf cleanly at the 2-inch height. Immediately follow up with a lowering to 1.5 inches for two mowings. Then, depending on maturity, the turf can be taken to 1 inch and then down to .75 inch. This depends, again, on maturity of the turf and the turf type. Some species such as the ryegrasses or bermudagrass are more tolerant to drastic mower reductions than bluegrasses or fine fescues. This is a judgment call by the superintendent based on the maturity of the turf and the environment. Each cutting height reduction, however, should include two mowings at that height before the next step down.

There are, without fail, additional shade and air circulation problems that become apparent during the later stages of grow-in or the first year's maintenance. Many owner representatives go to great lengths to ensure shade and air circulation problems are addressed during construction, but unexpected areas arise and the first year budget must allow for correcting some of these concerns.

This could be as simple as doing some additional underbrushing and raising of canopies in an area to improve air circulation. Maybe some select thinning and pruning of tree canopies to allow more morning sun penetration is all that is needed, or a fan added to improve air circulation in an area which can't be enhanced by natural means. Regardless of the situation, this is an area of turf maintenance that should be well explained to ownership so they are aware that the first year's maintenance is in large part a time where the intricacies of the golf course are learned by the superintendent, and then maintenance operations are adjusted and detailed to accommodate these intricacies.

For example, 'trouble greens' or disease indicator greens are usually apparent during the first year's maintenance. Very often, at least one of these is a different green from the one expected. The

Figure 9.5. Some tree attrition after construction is to be expected.

same may be true with a problem tee box or an area of the fairway that lacks density. There are also the one or two bunkers that always wash the worst after a rain and the areas of slower drainage always take time to find. These are just a few of many examples that must be taken into consideration with the first year's maintenance, and explain why budget flexibility must be planned for, so correction and fine-tuning can be accomplished.

Figure 9.5 is a common problem experienced in the first year after construction. Tree damage was experienced during the construction phase but did not show itself until the first year after construction.

Deep root feeding trees after construction is important to encourage root regrowth after reduction in the active root mass or construction shock. Deep root feeding specimen trees is best done during grow-in so shock recovery is promoted as soon as possible. This operation is usually not begun until the first winter season after grow-in because of time constraints and other priorities. Water soluble and slow release nitrogen fertilizers as well as a balance of other needed nutrients are important for regeneration of the root system in the protected areas after root pruning. To ensure the proper fertility mix to inject, take soil samples in the protected area and then consult with a knowledgeable arborist on rates, materials, and mixtures selected.

A typical deep root injection program uses liquid or soluble slow release fertilizers with minor packages according to soil tests and/or species needs. This is usually injected on a one-year fertilization program with injection sites at 3-feet intervals and about 8 to 12 inches deep. These injections are on a grid centered around the drip line of the tree.

The fertility rate is about 3 lb. of N/1,000 sq. ft. delivered in 50 gal. of water. If, for example, a 28-9-6 was being used, this would require 5.6 lb. of fertilizer per 50 gal. of water to deliver the desired 3 lb. of N/1,000 sq. ft. rate. The deep root feeder calibration is determined as time per injection hole.

1. Use 140 to 150 psi with the sprayer.
2. Time water collected in one minute at target psi from deep root feeder.
3. Determine the rate of solution needed per 1,000 sq. ft. and the number of injection sites per 1,000 sq. ft.
4. Determine the time in seconds to deliver the desired rate of solution per 1,000 sq. ft.
5. Divide the time per 1,000 sq. ft. by the number of injection holes per 1,000 sq. ft. to establish the injection time per hole.

Varying the injection amounts by as much as 20% is acceptable. The recommended injection pattern is diagrammed in Appendix 20.

Even with the best protection plans during construction, some tree attrition will occur. Sometimes it occurs from construction damage, but the problem usually persists from traffic underneath the drip line of trees. Chapter 3 details some tree concerns and needed protection management programs during the construction phase.

The first year maintenance budget requires some planning for

Figure 9.6. Wind damage can occur from 'sudden' direct exposure.

tree removal which can be costly on an established golf course. Removal of the tree and grinding the stump should all be accomplished at the same time so the damaged area left can be backfilled and reestablished with turf or a replacement tree. Replacement or the planned addition of new trees is a program that can begin successfully in the first year maintenance. This should be limited to previously identified key tree additions or replacements so minimal new tree care is added during a time when many other types of operations are underway.

Another key area where tree removal must be factored into the first year budget is unexpected tree damage from a wind storm. Figure 9.6 indicates the damage that occurs on new golf courses, especially ones carved out of a heavily forested area. Trees that were deep within a forest never received direct exposure to wind damage. Once the hole corridor has been cleared and graded, suddenly trees that were once in the middle of a forest are now experiencing direct wind contact. If tree trunks have any weak spots such as galls, this wind damage can cause breakage at that point which previously had not been a problem because of the protection from the forest.

This must be anticipated for two reasons. One is simply the need for a budget allowance to handle this expense. More importantly however, tree loss after completion of the golf course is a very difficult thing to accept by anyone, but especially by owners and developers because they do not understand the natural attrition rate that is going to occur from construction. Therefore, the owner representative should make owners aware during construction about tree loss that will occur then and the first two years after because of this natural attrition. When the first year budget is discussed in the early grow-in phases for planning purposes, this scenario can be explained so owners will be understanding. Consequently, when these types of problems occur, everyone is better prepared to deal with the removal and/or replacement of such trees.

CHAPTER TEN

PARTICULARS OF RENOVATING AN EXISTING COURSE

Renovation

When a decision is made for an existing course to undergo major renovation, many factors must be considered and researched to facilitate a smooth renovation in a timely manner. When a golf course is renovated by a nongolf course contractor with no experience, major damage to the unrenovated portions of the golf course can easily occur.

There are numerous details in major renovation that an experienced golf course contractor would have well planned out. Some of these would include:

1. how to locate staging areas around the course
2. where the spoil from green coring will be disposed of/utilized
3. the policy for use of the course rest rooms by construction personnel

4. construction traffic flow on the course that is least damaging
5. work ethics around golfers if course remains in play
6. interaction with superintendent as the crew maintains the course.

Redesigning and rebuilding of putting greens is important but just as critical is getting all the equipment and materials to the green sites with minimal damage. Moreover, the actual putting green complex must be rebuilt in an orderly fashion to keep the *destruction* confined to the site itself. Many golf courses have undergone renovation only to find cart paths, parking lots, and even some fairways destroyed because of poor construction traffic management during the renovation. Renovation is a very specialized business and is quite different and more difficult, in many respects, than new construction.

Prior to the beginning of renovation, haul roads must be established to accommodate construction traffic of equipment and materials around the golf course. This is very important for two reasons:

1. to facilitate construction traffic with as little damage as possible to the overall golf course, and
2. to facilitate construction traffic to allow the course to remain in play, if that is the goal.

Once travel routes are developed to minimize damage or exposure to golfers, the next step is to identify staging areas on the golf course. Staging areas are locations where material such as gravel, sand, and pipe are brought in and stockpiled to be distributed later by the contractor to the various site locations on the course. It is not hard to see that an inconvenient staging area can add costly travel time and expense to the construction project. Staging areas are centrally located or multiple in nature to better facilitate material distribution. They must also be accessible from highways to get raw materials to the site for redistribution.

Destruction of cart paths can be of major concern during ren-

Figure 10.1. Construction traffic can cause significant cart path damage.

ovation (Figure 10.1). There will be some degree of damage simply because the construction traffic will have to cross paths at some locations. However, this damage can be minimal. If paths are being replaced, the old paths can support the traffic so turf damage is minimized. This decision is made based on the overall renovation project but is important planning because a "haul road repair" budget line item must be incorporated into the master plan. Root pruning should also be assessed. Figure 10.2 illustrates the excessive root development occurring <u>under</u> paths in this renovation that is removed before new paths are put down.

Cart path relocation should also be carefully evaluated as part of the master plan. Often courses find that sections of cart paths are not used by golfers because of natural traffic flow or course design. The 'tendency' of traffic flow is the most direct route and golfer utilization of cart paths is poor if they are not located in natural flow corridors. Superintendents can be faced with a need for stakes/ ropes and directional signs, and all the maintenance headaches these can create, when better path location would have been the answer. Renovation can solve many of these traffic problems.

Figure 10.2. Root damage to cart paths can be significant.

Bunker Renovation

After bunker renovation it can be difficult to restabilize the edges and banks for sodding. Two methods have been used with good success and the choice usually depends on the severity of slopes. If the bunker is relatively flat, a plywood border about 6″ wide is installed at the bank/sand cavity interface. The flexibility of plywood allows it to be easily contoured with the bunker outline. It provides support to the bank during establishment and can later be removed after establishment.

The other method provides excellent stabilization for bunkers with steeper features including mounds, faces, or capes (Figure 10.3). This method utilizes burlap bags filled about three-quarters full of topsoil. The bags are laid in place and shaped to conform to the bunker edge contours. Soil is also filled into voids between bags if needed. The burlap soil bags can also be stacked to build up and stabilize a bank or cape in the bunker. The sod is then laid directly on the bags and the burlap will decompose over time, leaving a mature and stabilized slope/edge.

224

Figure 10.3. Successful bunker renovation requires quality construction.

Sod stacking or revetting is a design feature sometimes speced by the architect for bank stabilization. The architect will provide details on installation of the sod wall referencing stabilization, wall height, and slope steepness. One key to successful sod stack wall establishment is not trying to mow the wall face too quickly. Allowing the turf to grow and letting the face become "shaggy" produces better rooting within the wall and better stability. Later, the bunker face can be groomed, but mowing less than about 3 to 4 inches is not recommended for most turf species.

A major question asked when considering renovation is "Can the golf course be kept in play?" This is done predominantly by establishing temporary putting greens in front of existing green complexes. If tees are also being rebuilt, then keeping the course in play is usually impossible; however, there have even been temporary tees established. To answer this question of play during renovation, first determine the extensiveness of the renovation. If greens, tees, bunkers and irrigation are all to be updated and renovated, very common in older courses, then it will be virtually impossible to keep the course in play.

225

Golf courses often renovate only nine holes at a time so the other nine can remain in play. This is a possible alternative and the management must decide whether this is the best approach politically and monetarily for the course. However, renovating the entire golf course at one time is the most cost-effective and adds consistency to the project. Over the duration of the project it inconveniences the membership the least amount of time. If the golf course is to be done in stages, then it is totally inefficient to do less than nine holes at a time. Material consistency will also be lost by extending this renovation period over three or four years.

Some golf courses have created an interesting challenge for temporary play while greens are being rebuilt—playing the golf course backward with a tee box becoming the green. This has worked surprisingly well and has been fun and quite a novelty for the golfing membership. The tees are set up in the green approaches where the temporary greens would normally have been located. Then a tee box is selected as the green and a cup cut into it. This is totally possible if green renovation is being done and might be considered in the overall planning as a novelty for the membership during the construction process.

Identification of trees creating shade or air circulation problems and their consequential removal/disposal should be a part of the long-range plans for golf course renovation (Figure 10.4). There will never be a better time to remove trees detrimental to turfgrass health, or remove a misplaced/poorly chosen variety that has outgrown its intended look and unfairly interferes with play.

Careful evaluation of these concerns should be done in the renovation process so that trees and stumps are properly removed and disposed of. It is costly to properly take down a tree on an existing course because of the potential damage to the surrounding area from the equipment involved. A budgeted line item of tree removal, stump grinding, filling the hole, and reestablishing turf should all be a part of the renovation process.

An overall master plan or long-range plan is critical for golf course renovation. Redesigning green complexes, bunkers and bunker location, or enlarging or moving tees are only a part of the overall planning process for course renovation.

Figure 10.4. Poorly located trees create additional problems as they mature.

The long-range plan should also include a master plan for landscaping which would include both new plantings, whether it be trees or ornamentals, and removal of existing trees that are creating problems as outlined above. An arborist or landscape architect is an excellent source of assistance when evaluating new tree selections because comprehending mature size and shape of the tree species is essential. As an example, a tree placed on the inside of a dogleg to accentuate the turn should be taller and less broad when mature than a shorter and wider species. A taller tree is more effective in preventing a golfer from "cutting the corner" on a dogleg than a shorter, wider canopy tree. Appendix 21 is a typical example of the need for specific tree selection. This is just one of many examples where tree size and shape is of critical importance. Other key factors and tree selection criteria should include hardiness in a given locale and whether it casts shade on specific areas— greens in particular. Would it create morning or afternoon shade? All of these factors should be well defined when the landscape master plan is incorporated in the overall long-range plan.

227

CHAPTER *ELEVEN*

SPECIALIZED AREAS
OF CONCERN

New Course Overseeding in the South

Many southern golf course projects require overseeding fairways and tees the first winter after grow-in. This is nearly always for aesthetics, possibly for an improved appearance for real estate sales the first winter, or for an anticipated spring opening the following year. First year overseeding should be discouraged, however, because of the detrimental effects to the immature bermudagrass. The exception to this is light overseeding with an intermediate or perennial ryegrass when needed for E&S control support because of a lack of permanent turf development. Spring transition of perennial ryegrass can be a significant detriment to bermudagrass if not properly managed. Add to this a not-fully-matured bermudagrass, and a potential setback is created at a time bermudagrass is most vulnerable in its growing cycle.

The first winter after grow-in is a very important time for cleanup and repairs on the course that remain after construction/

Figure 11.1. *Perimeter cleanup in roughs on new course.*

grow-in. The maintenance crew should be able to focus their efforts on maintenance facility completion, equipment maintenance and repair, and course projects/problem solving, including drainage, tree pruning and cleanup, and bunker completion (Figure 11.1).

Focusing efforts on overseed mowing the first year can take away from the completion of the course and attention to fine-tuning details. If overseeding must be done that first winter, then rates should be kept as light as possible to encourage more of a color contrast than a pure ryegrass playing surface. Then next spring the ryegrass should be aggressively removed through cultural programs and select fertilization so the bermudagrass greenup will not be impeded any more than possible.

Ponds

Pond management is a critical part of overall golf course management, both in the grow-in and in the maintenance modes of turf care. Runoff from fertilizer, chemical, and sediment are of special concerns during grow-in because of the potential detriment to ponds, lakes, and streams.

When aerobic digestion, a pond's ability to clean itself up, is exceeded due to nutrient overload from surface runoff, the delicate balance of the pond is disturbed. In a normal pond, aerobic bacteria metabolize organic nutrients, making them unavailable for vegetative growth. This natural balance prevents algae blooms or other excessive vegetative growth in the pond. Nutrient overload throws a lake out of balance because bacteria cannot keep pace with the nutrient levels. Consequently, excessive vegetation develops very quickly.

Several factors can be designed into a lake to reduce potential vegetative buildup. The most prominent is to limit the amount of shallow water in a lake edge which allows sunlight penetration to the bottom for increased vegetative establishment. A 1:1 slope is recommended from a pond management standpoint to reduce the amounts of shallow water present, but a safety ledge usually 3 to 4 ft. wide is required in pond design to protect golfers and children. This safety ledge also serves as a mitigation area for wetlands and so this shelf area may be wider to increase square footage. Another form of safety ledge at lake edges is the littoral shelf described in Chapter 7.

Proper water aeration in a pond or lake is another critical aspect of maintaining a healthy aquatic environment. It is recommended that water in a shallow pond be turned over 4 to 7 times per day. This will allow the dissolved oxygen to stay in the 3 to 5 parts per million range which is adequate for a healthy aquatic environment. Spray fountains do not always provide sufficient pond aeration or are not necessarily the best means of aeration. As a general rule, 6 feet deep or greater is recommended for pond depth, with 8 feet being best. With deeper water it is easier to control temperature and light. Both these factors are important aspects of healthy aquatic environments. Detailed irrigation and fertilization management keeps nutrients from reaching ponds at unacceptable levels. This is the most basic approach to preventing nutrient overload and is part of the foundation programs for BMPs.

Nitrogen, particularly nitrate, and phosphorous are the two nutrients that are most commonly accumulated. Phosphorous plays a significant role in producing algae blooms which can be very detrimental to fish and other organisms by upsetting aquatic system ecology. Phosphorous usually moves with sediment, and research has shown that sediment movement is extremely low in established turf. However, in construction and grow-in this can be quite different. Sediment buildup can occur at a rate of 2 inches per year in the right circumstance—the equivalent of losing 80,000 gallons per service acre of storage capacity per year. Not only is poor sediment control damaging to the environment, but it can ruin the irrigation potential of a lake or pond very quickly.

A *no fertilization* zone of 15 to 25 feet around all pond perimeters has been recommended by many environmental concerns. Buffer strips can help in controlled runoff but cutting height, width, or aerification do not significantly affect the amount of nutrients in runoff, according to research.

Many concerns must be addressed about the infrastructure of ponds on the golf course property in the preplanning stages. Many development projects have used small ponds on the property to increase lot value for nongolf course lots. They have also been utilized in the course design considering playability and property values.

Ponds in and around a golf course have many functions beyond providing a source of irrigation water. When properly designed and built, ponds play a tremendous role in the Audubon Cooperative Sanctuary Program through the enhancement of many forms of wildlife.

Ponds also provide a critical function in the filtration of runoff water from the golf course and other adjacent areas. The effects of ponds to filter runoff water depend on the design aspects of that pond. The environmental consultant used to help with wetlands permitting should also be able to provide some details for pond construction to maximize wildlife habitat and riparian enhancement.

In the planning stages the superintendent should ask questions about golf course pond construction with three major points of concern:

1. the capacity of each individual pond
2. whether the ponds are interconnected for improved internal storage
3. if ponds are not interconnected, which ones are connected to the irrigation pond for increased storage.

Each pond should have a total volume calculation of water storage at normal fill level so the superintendent knows the ability of these ponds to supply irrigation water with respect to the water use study that should be completed. After the water use study is developed, oftentimes corrections to the master plan can be made by digging ponds with a greater capacity than originally designed to accommodate additional needed irrigation storage.

Sediment collection in ponds during construction and grow-in can accumulate to an unacceptable level and removal of this sediment is often attempted after grow-in is complete. Removal of sediment from a pond can be an expensive and difficult job, depending on the nature of the soil. The other facet of sediment removal is what to do with the material after it has been removed.

As part of the wetlands permit and/or the erosion and sediment control permit, the grow-in manager should question his options for silt removal if sediment buildup becomes a problem from heavy rains or other unanticipated events.

Many ponds serve as sediment retention basins during the construction and grow-in phase, and are often referred to as storm water management ponds (Figure 11.2). Selected vegetation around the edges is important for pond stabilization, but the long-term use of this pond as a permanent structure is important in the design and the size of the pond. During the initial phase of construction and grow-in, storm water management ponds often will have a sediment forebay construction site where biologs are installed at the water surface to further filter sediment out of the water at the point of sediment inflow. This confines the sediment

Figure 11.2. Interior ponds may serve as sediment basins during construction/ grow-in.

buildup to a small portion of the pond which can be more easily and cost-effectively cleaned out, restoring the entire pond to a much improved appearance for the life of the pond (Figure 11.3).

Potential groundwater contamination is another serious aspect when evaluating irrigation and fertilization practices. Specifications are very important and must be well planned by environmental managers if groundwater contamination is a specific concern. Details of these specifications will often be included in the permit (30). Properties that influence leaching potential of nutrient and chemicals include:

- Soil properties—porosity, permeability, and physical properties. Soils with a high CEC can hold or bind materials to a much greater efficiency. The depth of groundwater also plays a major role.
- Chemical properties—solubility, mobility, and persistence. Point source contamination which occurs when a chemical or nutrient enters groundwater by

Figure 11.3. *Lake/creek banks must be continually protected from erosion.*

misuse might include: spill, improper mixing, or improper loading. Nonpoint contamination occurs when a chemical or nutrient enters the groundwater by leaching or runoff.

One guideline for water usage according to environmental aspects of water quality addresses the effects that irrigation withdrawal could have on surrounding groundwater levels. This guideline states that there should be sufficient water available to meet irrigation needs without causing either a reduction of more than 5% of the 7-day, 10-year low flow level *or* substantially reducing the yield of existing wells in a given area.

Laws, bills, and acts governing water quality and use on golf courses include: Endangered Species Act, Clean Water Act, The Coastal Zone Management Act, The Safe Drinking Water Act, and The Federal Insecticide, Fungicide, Rodenticide Act. Some states have additional laws to regulate water quality and usage. Not all of these governing agencies apply to all water usage, depending on local and individual site conditions.

CHAPTER TWELVE

CLOSING COMMENTS

Sports Fields

In closing, a mention about sports fields should be noted if the golf course superintendent is ever responsible for a sports field complex. Sometimes sports fields such as grass tennis courts or croquet courts are a part of the country club complex. It is recommended regardless of grass type or maintenance level that the same practices outlined for appropriate golf course areas also be utilized on sports fields. Construction principles such as drainage, irrigation, sand root zones, grass selection and establishment are the same for intensely managed sports fields as the appropriate golf course counterpart (Figure 12.1). Basically the same mowing, watering, establishment, and cultural practices for fairways would apply. Fertility level, mowing heights, etc. can be adjusted according to playing conditions and the desired level of maintenance; i.e., a croquet court similar to a putting green; a football field similar to a fairway.

Utilizing the appropriate guidelines will produce the best playing surfaces in quality, durability, and ease of maintenance. Special note of equipment needs should also be researched from

Figure 12.1. *Sports fields have turf maintenance programs similar to corresponding golf course areas.*

Appendix 8 because most sports complexes have inadequate equipment.

In summary, this book has been designed to provide a reference to golf course superintendents for construction, renovation, and grow-in. It contains many details not found in any other publication to date. It is a collaboration of much field experience from the author and other respected experts in the field.

One of the major goals for me through this book was to provide as "hands-on" a resource as possible. My desire would be for this book to be on every turf manager's shelf, soiled and well-worn, as a useful field manual by the owner—the intention all along!

APPENDIX 1

Example Final Critical Path

Assume: 1. *Upper transition zone*
2. *Summer sprigging warm season grasses*
3. *Fall seeding cool season grasses*

AUGUST

*1. Wrap-up construction walk-through with management team to finalize completion of punch list, and address any issues to be completed

SEPTEMBER

*1. Finish seeding greens/deep rough areas, if applicable
*2. Finish bunker edging; spoil used for wash repair
3. Excavate bunker cavities, clean drains, level sand
4. Topdress low areas, light washes
5. Fill, patch, and/or sod deep washes
6. Lightly topdress tees
*7. Lime entire course; again, if needed according to soil tests
8. Fertilize/spike weak areas
*9. Fall fertilization—entire course
10. Slice/sweep fairways, roughs—warm season grasses
11. Mark contour mowing patterns—fairways
12. Finish level/set of irrigation heads
13. Cart path turnouts; repair fill behind curbs if needed
14. Cart path road crossings/repairs
15. Repair bridges
*16. Grow-in bentgrass greens
17. Begin complete rough cleanup, mulching

18. Hydroseed tall fescue—deep roughs
19. Repair/flush all drainage systems
20. Hydroseed wildflower areas
*21. Ensure punch list is complete

OCTOBER

*1. Grow-in greens
2. Sweep fairways—cleanup/overseeding prep
3. Fertilize entire course for winter
*4. Overseed tees, fairways, if applicable
*5. Continue rough cleanup, mulch, seeding
6. Complete bunker cavities, sand spreading
7. Complete washout, topdressing
8. Grow-in of overseeding
9. Fill in and repair cart path edges
10. Sweep/wash cart paths
11. Measure golf course
12. Begin finish details of maintenance facility
13. Confirm start of rest station/rain shelter underground services

NOVEMBER

1. Continue rough cleanup, mulching, rough grow-in
*2. Begin setting down all cutting heights and mowing patterns
3. Firm up bunker sand
4. Begin ornamental plantings on course
5. Feed all key trees—deep root feeding
*6. Fertilize overseeding, if applicable
*7. Organize maintenance area/crew training
8. Budgets for upcoming year
9. Begin construction of rest stations/rain shelters
10. Maintenance facility completed
* 11. Follow up soil tests from preplant
12. Adjust critical path according to schedule flow for winter cleanup

DECEMBER

 *1. Major PM on equipment
  2. Mowing—tees, fairways, greens (if applicable)
 *3. Rough/perimeter cleanup
  4. Begin tree planting on course
  5. Continued maintenance area organization
 *6. Apply needed soil amendments according to November soil tests

JANUARY/FEBRUARY

  1. Rough cleanup continued
  2. Mowing greens, deep roughs as needed
 *3. Fertilize overseeded areas, fescue in roughs
 *4. Construction of fertilizer/chemical/seed building and soil, sand, mulch bins
  5. Continued tree planting
  6. Tree pruning/stump grinding
  7. PM on all equipment—continued
  8. Spot winter annual weed control

MARCH

  1. Additional rough mulching, reseeding fescue areas as needed
 *2. Slice/fertilize entire course
 *3. Mid-March—aerate, vertical mow, topdress greens, collars (removing grow-in layering)
 *4. Brush overseeding, grooming
  5. Repair any inoperable drainage/add additional as needed
  6. Add any additional trees/ornamentals desired
  7. Add any additional color on course desired
 *8. Complete rain shelters/rest stations
  9. Evaluate preemergent application for summer annuals based on turf maturity

APRIL

*1. Fertilize lightly April 1 for maximum color at opening- if spring

*2. Burn in mower stripping patterns

3. Overall cleanup wherever needed

4. Sweep, edge cart paths

5. Fertilize entire course

6. Brush/slice/vertical mow all overseeded areas

*7. Set all golf course accessories—distance markers, ball washer posts, benches

*8. Aggressive grooming of greens for smoothness and density; finalization of maturity

9. Removal of silt berms on cart paths and appropriate drainage additions

MAY

*1. Walk-through with management team and architect to confirm attention to all details—design/playabilty are main focus at this point

*2. Provide the director of golf with cultural schedule (aerification, topdress) for upcoming summer so exact needed schedule can be implemented

---

* Priority items

Appendix 2

Architect Drawings Are
Excellent Field Notebooks

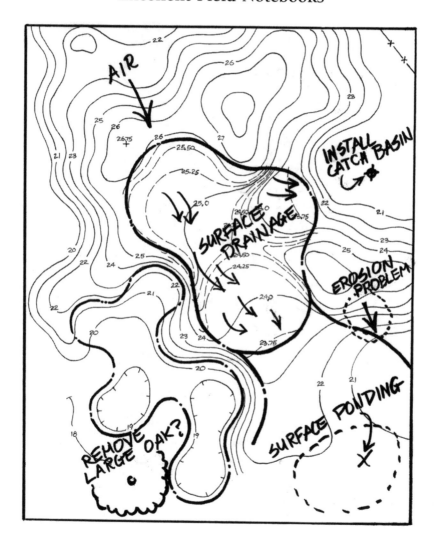

APPENDIX 3

Properly Detailed Asbuilt Drawing

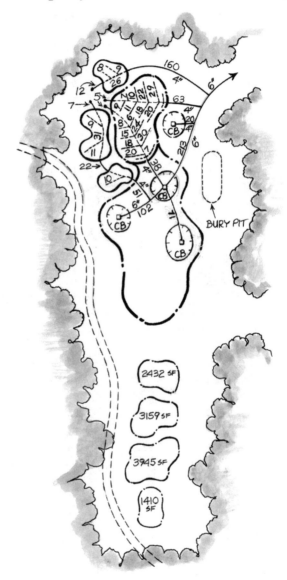

APPENDIX 4

Average Projected Water Availability vs. Need

Average Projected Available Water Balance for Irrigation Lakes

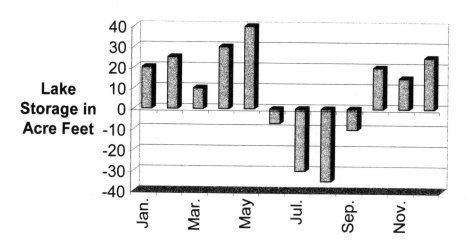

APPENDIX 5

Water Suitability Standards for Irrigation

Turf Irrigation Water Guidelines for Suitability

Water Problem	Measurement	Units	Water Suitability		
			No irrigation restriction if:	Slight to moderate restriction if:	Severe irrigation restriction if:
Salinity	Electrical conductivity (EC)	Decisiemens per meter (dS m^{-1})	Less than 0.7	0.7 to 3	Greater than 3
	Total dissolved solids (TDS)	Milligrams per liter (mg L^{-1})	Less than 450	450 to 2,000	Greater than 2,000
Soil water infiltration	Electrical conductivity (EC) and sodium adsorption ratio (SAR)	SAR=0 to 3 and dS m^{-1}=	Greater than 0.7	0.7 to 0.2	Less than 0.2
		SAR=3 to 6 and dS m^{-1}=	Greater than 1.2	1.2 to 0.3	Less than 0.3
		SAR=6 to 12 and dS m^{-1}=	Greater than 1.9	1.9 to 0.5	Less than 0.5
		SAR=12 to 20 and dS m^{-1}=	Greater than 2.9	2.9 to 1.3	Less than 1.3
		SAR=20 to 40 and dS m^{-1}=	Greater than 5	5 to 2.9	Less than 2.9
Sodium ion (Na) toxicity in roots		SAR	Less than 3	3 to 9	Greater than 9
Sodium toxicity in leaves		milliequivalents per liter (meq L^{-1})	Less than 3	Greater than 3	
		milligrams per liter (mg L^{-1})	Less than 70	Greater than 70	
Chloride ion (Cl) toxicity in roots		meq L^{-1}	Less than 2	2 to 10	Greater than 10
		mg L^{-1}	Less than 70	70 to 355	Greater than 355
Chloride toxicity in leaves		meq L^{-1}	Less than 3	Greater than 3	
		mg L^{-1}	Less than 100	Greater than 100	
Boron ion (B) toxicity		mg L^{-1}	Less than 1	1 to 2	Greater than 2
Bicarbonate		meq L^{-1}	Less than 1.5	1.5 to 8.5	Greater than 8.5
		mg L^{-1}	Less than 90	90 to 500	Greater than 500
Residual chlorine		mg L^{-1}	Less than 1	1 to 5	Greater than 5

Reference source #29

APPENDIX 6

Dual Irrigation System for Green Complex

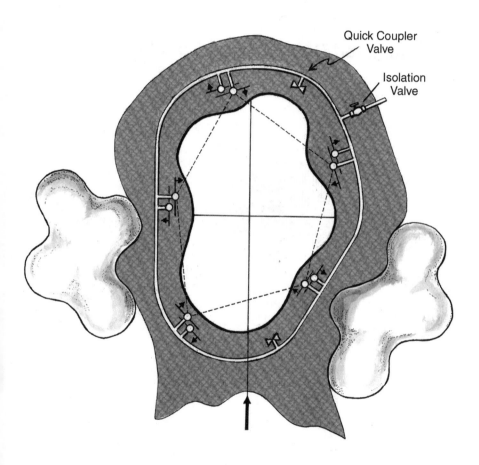

APPENDIX 7

Estimated Golf Course Construction Breakdown

Estimated Construction Cost Breakdown
Note: Grow-in not identified

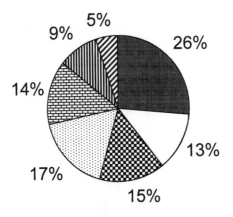

Irrigation/pump
Green/tee construction
Cart paths
Fwy tile/grassing
Earthwork/shaping
Drainage/lake construction
Taxes/bonds

Source:GCBAA Guide to Estimating Cost for Golf Course Construction

APPENDIX 8

Suggested Budgetary Need Breakdowns
Grow-In Budget

The grow-in budget is quite different from maintenance due to variations in spending areas. Labor, as with the maintenance budget, is the greatest expense, but fertilizer is always the largest material expense.

Normally, a grow-in budget should be scheduled for a six-month period beginning when planting starts. It is presented as a separate budget from the first year maintenance budget. The grow-in budget is put together in the planning stages along with the following other budgets:

1. Grow-in equipment
2. Equipment
3. Maintenance facility
4. Maintenance facility stocking
5. Mechanic shop
6. Initial course setup
7. First year maintenance

These budgets are critical for developers to be aware of so planning and budgeting the "entire" construction/grow-in phases can be accomplished. Many projects do not have these areas outlined and priced; consequently, they oftentimes are significantly undercapitalized because by the time grow-in arrives, funds are tight.

Following is a suggested line item list of budget areas and supporting notes:

1. Labor—should include assistant, mechanic, irrigation technician, and full-time/part-time employees—6 month duration
2. Soil amendments (usually needed about 8 to 10 months after preplant materials, based on follow-up soil tests)
3. Fertilizers—granular, liquid, and/or fertigation
4. Chemicals—plan for weed control on spot treatment basis after maturity; fall armyworm control; fungicide preventative to greens
5. Additional sod—30,000 sq.ft. is a good estimate
6. Erosion and sediment control—silt fence, mulch, jute mat, etc., according to E&S plan
7. Bunker sand specs/trim/clean drains/spread sand—contractor's responsibility??
8. Topdressing—greens, collars, tees and approaches, ideally
9. Ryegrass—will you need for winter stabilization? (200 lb/ac average)
10. Seed—e.g., lovegrass, wildflowers, roughs—repair and touch up
11. Washout repair—soil, sand, gravel
12. Additional drainage—late surfacing wet springs or settled areas—sand, gravel, perforated pipe
13. Irrigation head leveling—is it included in contractor's job??
14. Gas/oil
15. Repairs
16. Soil testing
17. Contingency—planning for unexpected (5% of budget)
18. Equipment rental—may be necessary depending on your site availability (e.g., backhoe, trencher)
19. Tree removal—some dying trees and some shade/air circulation corrections/stump grinding
20. Ornamentals—as part of the master landscape plan (refer to first year budget)

Minimal Equipment List for Grow-In

Quantity	Type
2-3	*84″ belt driven triplex
5-6	Used walk greens mowers
1	Spray unit-150-200 gal- w/boom, walk boom and hose reel
2	Tractors
1	Vicon spreader
1	Triplex
2	Flymos
1	Rotary (72″)
3	Walk spreaders—rotary
2	Walk spreaders—drop w/agitators
2	Topdressers
2	Drag mats
1	Drag brush
1	Pickup truck
2	Heavy utility vehicles
4	Utility/transport
1	5 Gang pull unit
1	Small trailer
4	Greens mower trailers
1	Dump trailer
1	Pulverizer
1	Backhoe unit
1	Small dump truck
1	Wire locator
3	Hose/syringe outfits
-	Radio system
-	Misc. irrigation tools
-	Misc. tools
-	Irrigation supplies

* Some mower inventories vary depending on course size

Example Golf Course Equipment Inventory
**Average 18 Hole Operation

GREENS

*2	Triplex/5 walk mowers
2	Walk mowers—cleanup, collars
4	Used walk mowers—grow-in
1	Topdresser—tow type
1	Topdresser—walk-behind
2	Walk-behind deep vertical mowers
2	Greens mower trailers
*2	Greens mower trailers
-	Aeration equipment—selection optional
2	Power spikers
2	Drag mats—3 × 5′ and 6 × 5′
1	Drag brush
1	Set vertical mow units—triplex
5	Whip pole

* Optional

TEES

2	Triplex
2	Walk mowers—grow-in conversions
1	Greens mower trailer

FAIRWAYS

2	Belt driven triplexes—grow-in (slope conversion)
1	5 gang lift unit
1	1400 lb. Vicon
1	Slicer/aerifier
1	Sweeper-rotary brush
1	PTO blower
1	Sweeper/vac
1	Small complex mower (triplex)

ROUGHS

1	5 gang lift
1	Trim mower—slopes (triplex)
2	Out front rotaries

OTHER (GENERAL USE)

1	Backhoe (used?)
1	Dump truck (used?)
1	Dump trailer—bunkers/topdressing
2	Tractor- 40-45 Hp
1	Tractor- 33-35 Hp
1	200 gallon spray unit with spray hawk, hose reel
1	300 gallon spray unit with boom
2	Rotary mowers
3	Flymos
2	Edgers (or string-head trimmers—bunker edging)
1	Bunker rake—with blade
1	Bush hog
2	Multiple use heavy utility vehicles
2	Heavy duty utility vehicles
6	Utility vehicles—lightweight transports
1	Pickup
2	Utility trailers—8 × 12'
1	Utility vehicle—irrigation
-	Hand tools (see itemized list)
1	Smoothing box—rollover
1	Pulverizer
1	Brush chipper
1	Level/transit/rod
2	Rollers- 24" and 36"
-	Syringe equipment—hose, reels, nozzles, etc.
-	Irrigation equipment—set heads, couplers, etc.

-	Golf supplies (see itemized list)
**1	10' drop spreader
-	Power tools (see itemized list)
2	Walk blowers
1	Wire locator
1	Metal detector
1	Trash pump
1	Fault detector
1	Backup reel unit—for each mower
1	Sod cutter- 18"
2	Pluggers 2" and 6" each
7	Radios
1	Utility vehicle vicon
1	Fuerst harrow
**	If overseed fairways

** These equipment recomendations are to be used as guidelines. Needs vary according to design, topography, turf selections, and budgets.

Mechanic Shop Inventory

Average 18 Hole Operation

1	Pallet jack—4,500 lb. capacity
1	Reel spin grinder
1	Bedknife grinder
1	Shop fan on stand
1	Hydraulic tester
2	Jumper cables
1	Hand truck
2	Bench vise—4" and 8"
1	Air compressor—5 hp 60 gal.
1	Portable air tank
1	Master mechanic tool set
1	Small tool set—field
1	Bolt cutter—24 in.
1	Torque wrench

1	Battery charger/tester/starter
1	Floor creeper/creeper seat
4	Jack stands
2	Floor jacks—2 and 4 ton
1	Timing light
1	Electric load tester
1	Handheld tachometer
2	Shop lights—retractable
1	Puller/hoist
4	Extension cords- two 50′ and two 100′
20	Shelving Units—16 × 36 × 72″
2	Portable workbenchs—heavy duty
-	Built-in workbenches (allowance)
2	Lapping machines
2	50′ Air hoses
2	25′ Air hoses—coiling
2	Pneumatic wrenchs- 1/2″ and 3/8″ drive
1	Set pneumatic wrench tools
1	8 ft. ladder
1	Extension ladder—24′
3	Pipe wrenches—10″,14″, and 18″
1	MIG wirefeed welder
1	Oxyacetylene outfit/tank cart
-	Welding accessories
-	Deposit on O_2, ace., argon tanks
1	Tire changer—self changer—manual
4	Drum cradles
3	Drum pumps
1	Drum wrench
1	4′ Level
1	Wrecking bar
1	Drill press on stand—1/2 hp
1	Hand grinder (2 hp)—9″ disc
1	Small hand grinder—4″ disc
1	1/2″ Drill-varispeed, reversible
1	3/8″ Drill—varispeed, reversible
1	Drill bit set

1	Bearing press—H frame
1	Seal/bearing drive set
1	Bench grinder on stand
1	Circular saw (2 3/4 hp)
1	Shop vac 16 gal.
-	Nut/bolt assortment
-	Other fastener assortments
1	Solder gun set
1	Pressure washer 5 hp
1	Parts washer, 5 gal. (VAT)
1	18″ Accu-gage mower H.O.C.
1	Hydrometer
1	Reciprocating saw
5	Grease guns—pistol w/flex hose
1	Band saw 12″
1	Set C clamps
1	Vise grips—welding "C"
-	Hydraulic floor lift unit
-	Initial parts stock
-	Mechanic contingency (specialized tools)
1	Heavy chain—15′ and 25′ (or tow strap for

one)	1 Extra heavy chain—20′
-	Initial irrigation room repair stock—tools, parts, fittings, pipe, etc.

Example Course Supply Stock

Average 18 Hole Operation

I—Course

10	Ball washers
20	Benches
20	Trash receptacles
20	Hole signs
6	Water coolers with stands

40	Tee markers—each setup
27	Regulation cups (plastic or metal)
18	Practice green cups
27	Flagpoles
27	Flags
20	Practice green markers
60	Bunker rakes—average 2/bunker plus inventory
-	Distance markers/info signs
80	32" Rope stakes
40	16" Rope stakes
2	Rolls braided rope

II—Maintenance

2	Cup changers
5	Gallons ball wash soap
2	Cases tee towels
2	Cup pullers
2	Cup setters
2	Buckets/spades/screwdrivers
2	Divot repair tools

Hand Tool Stock
Average 18 Hole Operation

10	Round point shovels
10	Square point shovels
4	Grain scoops
2	Trench shovels—4"
3	Sharp shooters
2	Round point D handle shovels
2	Square point D handle shovels
8	Bow rakes—14"
9	Leaf rakes

4	Spring tooth rakes—42″
6	Aluminum rakes—24″
2	Aluminum rakes—36″
4	Roller squeegees—36″
5	Bank blades
2	Scuffle hoes
1	Concrete hoe
1	Mortar pan
1	Wheelbarrow
2	Sledge hammers—8 lb.
2	Axes
2	Sod lifters
2	Turf edgers
1	Pick
2	Mattocks
1	Hole digger
2	Pitchforks
2	Mulch forks
4	Level lawns
4	Plug pushers
1	Pipe probe
1	Heavy pry bar
1	Shoulder seeder
2	Pole saws
2	Bow saws
3	Pruning saws
2	Lopping shears
2	Pruning shears—hand
2	Hedge trimmers—hand
1	Gas hedge trimmer

Appendix 9

Maintenance Facility
Zones and Major Design Considerations

Administration Zone
   Superintendent office
   Assistant office
   Rest room & shower—men and women
   Drawing & map area
   Filing area
   Secretary work area
   Dormitory area for turf students
   Locker room—crew
   Central heat and air
   Lunch/break room with vending
   Kitchen facility
Consider: Access from parking and course, traffic flow

Equipment Maintenance & Repair Zone
–Repair
   Floor lifts—15 ft. ceiling minimum
   Lighting
   Work area for large equipment
   Air outlets
   110 & 220 power outlets
   Proper door sizing
   Separation of repair area (traffic flow)
   Floor drain (check codes)
   Wash sink
   Mechanic office

Workbench
Welding table
Steam cleaning allowance
-*Parts*
Security locked
Shelving
-*Lubrication*
Security locked
Fire protection (check code)
Ease of truck delivery
Oil & grease disposal
Ventilation

Equipment Storage Zone

Air, water, power outlets properly located
Floor drains
Wall protection—8 ft. high—wood
Overhead doors
Traffic flow
Covered equipment storage outside

Golf Course Supply Zone

Adequate shelving

Irrigation Zone

Secured area
Adequate shelving
Workbench
Air/power outlets
Proper lighting
Pipe rack—outside in covered storage

Tool Storage Zone

Secured area
Adequate wall space for hangers
Power tool storage

Seed Storage Zone

Rodent protection
Humidity controlled

Chemical Storage Zone

Self-contained unit preferable
Flood shower—emergency
Eyewash
Sink
Secured area
Proper signage
Ventilation fans
Explosion-proof lighting
Noncorrosion shelving/work table
Fireproof
Heat equipped
Delivery accessible
Insurance compliances checked

Fertilizer Storage Zone

Block construction
Truck access—loading dock
Ventilation
Humidity control
Explosion-proof lighting

Fuel Station Zone

Traffic flow
Security
Safety requirements—local codes

Aboveground tanks
Cement pad and post protection

Material Storage Zone
-Bunker sand, gravel, mulch
10 ft. × 20 ft. each
Open bins
Poured slab
Wall protection—5–6 ft. high
-Topdress
50 ton capacity
Covered bin
Wall protection
Accessibility for dump trucks & loader

Wash/Loading Zone
2″ water line for sprayer fill
Pressure/steam wash
Air outlet
Proper drain—recycle system
Traps—grass clippings, grease
Sprinkler/valve test connections
Codes met for fill station, i e., curb containment,
covered area

Trash Collection Zone
Dumpster
Accessibility for pickup

Parking
Traffic flow
Delivery access
Loading dock setup

Paint Room (optional)

Exhaust fan
Sealed area
Explosion-proof lighting
Suspended beams—paint item hanging
Electricial outlets & covers
Local codes/permits must be obtained

Maintenance Facility
Quality Control Punch List

- Check all building codes carefully. This must include all special requirements such as chemical storage, handicap specs, exit signs, hydrants, oil/water separator, sewer specs for wash area, etc.
- Specify door types and sizes
- Specify security system
- Adequate outside lighting—security and work conditions before daylight
- 200 amp electricial box minimum
- Adequate skylights
- Separate power line for computer
- Adequate telephone line requirements
- Properly located 110, 220 power, and air outlets throughout building
- Specify downspouts and tiled away from building
- Extra 4" PVC conduit speced in slab for future expansion
- Seal concrete slab with high quality sealant for cleanup and dust control
- Concrete skirts at least at doors, preferably entire perimeter
- Post traffic protectors at door corners, fuel area, and other pertinent locations
- Double check traffic flow and parking during site grading
- Line interior walls with 8 ft. plywood to protect metal siding from traffic
- Provide for possible expansion in future
- Maintenance facility site properly fenced; check gate locations and size carefully
- Include budget for landscape
- Specify building exterior coordinated with appearance of project
- Spec needs in administration area for furniture, carpet, trim, shelving, storage, etc.
- Verify adequate planning for shelving and hardware in all areas needed

Appendix 10

Example Daily Planner as Grow-in Manager

Early a.m.—*Ride course to check for stuck/weeping heads, blow-outs. Record water total for last 24 hours
*Disease/insect scouting
*Mowing schedules for the day
*Misc. schedule for the day; e.g., bunkers, cleanup, E & S maintenance
Fertilization for the day
Make note of birdbaths on your field asbuilts/hole diagrams

Late a.m.—*Do you see any dry areas?—ride the course looking closely for this. Check growth rate in tees, fairways, etc., from last mowing

Noon —*set assignments
—set p.m. watering

Early p.m.—Walk 6 holes minimum with field prints and camera—superintendent and assistant together. This allows time to discuss management scheduling and evaluate progress.

Mid p.m. —Mow greens during grow-in—dry initially makes an important difference. Determine water needs for afternoon; from 6-hole walk, make water schedule for next day

Later p.m.—*Set fertigation for night; check water again

*Especially important jobs to train assistant in monitoring/scheduling

**Set meeting schedule with management team *before* grow-in begins so you can establish monitoring routine.

Appendix 11

Catch Basin Installation Cross Section

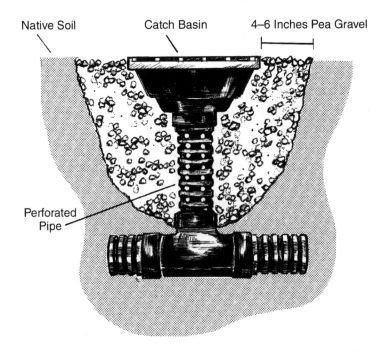

Native Soil

Catch Basin

4–6 Inches Pea Gravel

Perforated Pipe

APPENDIX 12

Green/Bunker Drainage Incorporation

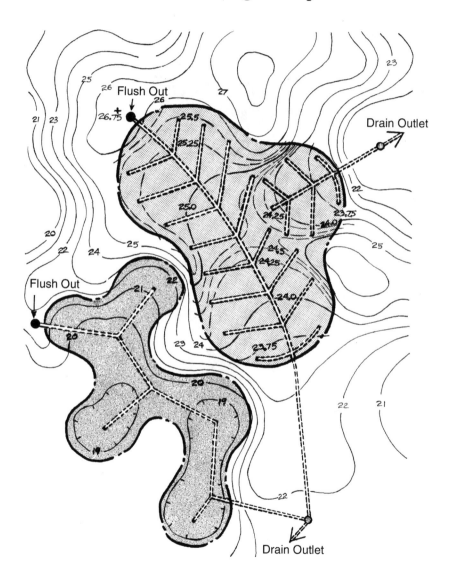

APPENDIX 13

Phosphorus Needs at Seeding

15–20 ppm sufficient for maintenance

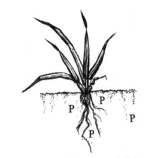

50–70 ppm during establishment

Appendix 14

Fertigation Calibration Worksheet

1. What is the size of the irrigated area?

The first question is what is the irrigated area? This is a figure that you or the irrigation designer will have. The area will be in acres or square feet. To do most nutrient calculations you will need to convert the area into 1,000 square foot units.

Example:
1 acre = 43,560 sq. ft.—43.56 units

Application Rate
Example:
4 oz. per 1,000 sq. ft./month of soil conditioner
1 acre application = 43,56 x 4 oz. = 174.24 oz.

If the irrigated area of the golf course is 70 acres:
174.24 x 70 = 12,197 oz.

Convert to gallons = divide by 128 oz.
12,197 ÷ 128 = 95 gal. of soil conditioner/month

When you want to plan a feeding program, the next question is how much bulk liquid nutrient do you want to apply?

Example:
You want to apply .25 lb. of (N) Nitrogen per 1,000 sq. ft. to the 70 acre golf course area:

25 lb. x 43.56 x 70 ac = 762 total lb. of N
You are using a liquid fertilizer with an analysis of 20-2-3.

Important
Dry and liquid nutrients are both calculated by weight.
The liquid 20-2-3 nutrient weighs 10.8 lb. per gal. which calculates:

(N) = 10.8 x .20 = 2.16 lb. per gal.
(P) = 10.8 x .02 = .22 lb. per gal.
(K) = 10.8 x .03 = .32 lb. per gal.

Calculate the amount of nutrient you want to apply:
Each gallon of 20-2-3 contains 2.16 lb. of N

To calculate gallons of nutrient needed:
Divide 762 lb. N needed by 2.16 lb. N per gal = 353 gal.
of 20-2-3 fertilizer to apply .25 lb. N to the 70 acres.

2. **How do I apply it?**
If you plan to apply the 95 gal. of soil conditioner monthly
you could use a micro pump (2.5 gph capacity). This would be
set using the following calculations.

Irrigation cycles per month ____Length of irrigation cycle
(approx.) _____ hr. Application amount _____gal.

Example:
20 cycles, 7 hr., 95 gal. to be applied
20 cycles x 7 hr. = 140 hr. total hr.

Application rate = Amount to be applied ÷ total hr.
of application
95 gal. ÷ 140 hr. = .68 gal. per hr. rate

**Pump stroke setting is determined by dividing the application
rate per hr. by the pump capacity**

The application rate is .68 gph. The micro pump capacity is 2.5 gph.
.68 gph ÷ 2.5 gph = .272 = 27% stroke setting

This same formula is used for larger capacities of fertilizer in the larger pumps

Target: 353 gal. per month of blended fertilizer (20-2-3)
Cycles: 20   Length of cycle: 7 hr.   Pump capacity: 27 gph
353 gal ÷ 140 hr. = 2.5 gal. per hour
2.5 gph ÷ 27 gph = .09 = 9% stroke setting

Set the stroke setting and mark the tank or drum. Run two to three cycles and check the tank or drum to verify the correct rate. You may have to increase or decrease slightly to make your exact target rate.

Injection Application Rate Worksheet

Nutrient rate per 1000 sq. ft. _____ is N1
Area irrigated _____ acres is A1
Pump capacity _____ gph is PC

Nutrient Volume (NV) per acre total
NV = N1 x 43.56 x _____ A1 = _____ in oz.:
　　　divide by 128 = _____ gal.

Total volume (TV) = A1 x NV in gal.

Irrigation run time _____ hr. is T1
Number of cycles _____ is C1.

Total hours of run time (TH) = T1 x C1

To find Application Rate (AR) = TV ÷ TH (total gph)

To find pump stroke setting divide pump capacity (PC) by AR

PC _____ ÷ AR = x 100 = _____% stroke setting

Calculations:

Summary Calculations

Compute nutrient budget
- Ounces per 1000 sq. ft. x 43.56 = oz. of product per acre
- oz./acre ÷ 128 = gal. per acre
- Acres x gal./ac. = total gal.

Compute application rate
- Irrigation run time x cycles = total run time
- Total gal. ÷ total run time = gph

Compute pump setting
- Gal. per hour ÷ pump capacity x 100% = stroke setting

APPENDIX 15

Grow-in Fertility Needs by Turf Type/Area

Bentgrass Greens—High Sand
Assume Late Summer/Early Fall Seeding

Days After Seeding	Suggested Ratio	Rate=lb./1000 sf
Preplant by soil test is incorporated into top 1"@ final floating	Starter	1.5 lb. P
	Granular micronutrient	by label
	Humate/Carbohydrate	by label
	Slowly Available N	1.25 lb. N
14	*23-4-10	5
19	18-0-18	4
24	15-0-30 plus Fe	4
29	23-4-10	5
34	18-0-18	4
39	23-4-10	4
44	18-0-18	5
49	23-4-10	4
54	0-0-28 plus Fe	4
59	23-4-10	4
64	18-0-18	4
69	15-0-30 plus Fe	4
80	23-4-10	5
100	18-0-18	5
120	18-0-18	4

Dec. dormant feed

Jan.—now
fertilizing for soil
nutrition & color as needed

* Assuming balanced soil fertility based on preplant soil testing. Analyses listed are suggested, and similar ratios based on availability will substitute. Rates may have to be altered to accommodate special needs. These may include salt problems, extremely low P or K or possibly additional Fe needed.

   # Now begin reference to the first calendar year's fertilization schedule.

Bentgrass Greens—High Sand

SPRING SEEDING

Days After Seeding	Suggested Ratio	Rate=lb./1,000 sf
Preplant by soil test is incorporated in top 1″ @ final floating	Starter	1.5 lb. P
	Granular Micronutrient	by label
	Humate/Carbohydrate	by label
	Slowly Available N	1.25 lb. N
14	*23-4-10	5
19	18-0-18	4
24	15-0-30 plus Fe	4
29	23-4-10	5
34	18-0-18	4
39	23-4-10	4
44	18-0-18	4
49	23-4-10	3-4#
54	0-0-28 plus Fe	4
59	23-4-10	3-4#
64	18-0-18	3-4#
69	0-0-28 plus Fe	5
80	15-0-30	4

* Assuming balanced soil fertility based on preplant soil testing. Ratios and rates may have to be altered to accommodate special needs. These may include salt problems, extremely low P or K, possibly additional Fe needed.

Rate depends on locale and spring temperatures. Less N preferred if summer temperatures start early.

Bentgrass is hardened off for summer as well as possible. Taller cutting height most important until fall. Summer color best maintained with solubles. Make 3 summer applications of 0-0-50 at .75 lb. K/1,000 sf. Color can be supplemented with soluble Fe or Mg.

about 180-190	12-25-15	4
205	18-0-18	4

** Refer to the first calendar year's fertilization schedule for suggestions. Must be modified for spring seeding and new soil tests.

Bentgrass Tees / Fairways

Days After Seeding	Suggested Ratio	Rate=lb./acre
Preplant by soil test is incorporated into 1"@ final floating	Starter *Slowly Available* N	1.5 lb. P 1.25 lb. N
14	26-3-6	130
21	26-3-6	100
30	19-0-19	120
40	27-5-10	125
60	10-20-20	200
Dormant feed	20-5-20 w/Fe	225
Early spring	18-25-12	200
30 days after starter	25-5-11	160

*Tees may require same preplant as outlined for greens.
**Begin first year maturity/maintenance fertilization. New soil tests are needed <u>before</u> early spring applications to finalize fertility programs.

Cool Season Tees / Fairways / Roughs

Days After Seeding	Suggested Ratio	Rate=lb./acre
Preplant by soil test is incorporated into 1"@ final floating	Starter *Slowly Available* N	1.5 lb. P 1.25 lb. N
14	32-3-9	130
21	32-3-9	130
30	15-0-15	130
40	27-5-12 w/Fe	125
60	19-0-19	150
Dormant feed	20-5-20 w/Fe	225
Early spring	18-25-12	200
30 days after starter	25-5-11	175

* For nonirrigated fairways and roughs, reduce rates 25%.

Begin first year maturity/maintenance fertilization. New soil tests are needed <u>before</u> early spring applications to finalize fertility programs.

Hybrid Bermudagrass Greens—High Sand

SPRIGGED

Days After Sprigging	Suggested Ratio	Rate=lb./1,000 sf
Preplant by soil test is incorporated into top 1" @ final floating	10-20-20 + 3% Mn *Slowly Available* N	11 1.25 lb. N
1	30-0-0 liquid	.20 lb N
7	15-3-15+1%Fe, 1%Mn, 1.5%Mg	
11	21-0-0	3.5
12	30-0-0 liquid	.20 lb N every other day
16	15-3-15+1%Fe, 1%Mn, 1.5%Mg	6.25
20	21-0-0	3.5
24	15-3-15+1%Fe, 1%Mn, 1.5%Mg	6.25
28	21-0-0	3.5
32	10-20-20+3% Mn	11
36	21-0-0	7.5
40	5-3-15+1%Fe, 1%Mn, 1.5%Mg	6.25
44	21-0-0	3.5
48	15-3-15+1%Fe, 1%Mn, 1.5%Mg	6.25
52	21-0-0	3.5
56	15-3-15+1%Fe, 1%Mn, 1.5%Mg	6.25
60	21-0-0	3.5

If Sprigging Late Summer:
Same basic program as weather allows; however, potassium must be kept high to harden plant as it is going into fall more immature.

Hybrid Bermudagrass Tees / Fairways / Roughs
SANDY SOILS

Days After Sprigging	Suggested Ratio	Rate=lb./acre
Preplant by soil test is incorporated into top 1″ @ final floating	10-20-20+3% Mn *Slowly Available* N	250 50 lb. N
10	15-3-15+1%Fe, 1%Mn, 1%Mg	300
17	21-0-0	180
24	15-3-15+1%Fe, 1%Mn, 1%Mg	300
31	21-0-0	200
38	15-3-15+1%Fe, 1%Mn, 1%Mg	300
45	10-20-20 + 3% Mn	250
52	21-0-0	200
59	15-3-15+1%Fe, 1%Mn, 1%Mg	300
69	21-0-0	200
76	15-3-15+1%Fe, 1%Mn, 1%Mg	300
86	21-0-0	200
93	15-3-15+1%Fe, 1%Mn, 1%Mg	300

* For nonirrigated roughs, reduce rates accordingly

OR

** Reduce rates by 1/3 for Piedmont/mid-South.

If a fertigation system is to be used, operate every other day for the first 14 days at .20 lb. N/1,000 sf. Then continue at same rate every third day for the balance of grow-in.

If Sprigging Late Summer:
Same basic program as weather allows; however, potassium must be kept high to harden the plant as it goes into fall more immature.

Seeded Bermudagrass Fairways / Roughs

ASSUME EARLY SUMMER SEEDING

Days After Seeding	Suggested Ratio	Rate=lb./acre
Preplant by soil test is incorporated into 1"@ final floating	Starter *Slowly Available*	65 lb. P 50 lb. N
21	34-3-8	100
31	18-0-18	200
40	21-0-0	200

Postemergent spraying can usually begin about now if three or more mowings have been done.

50	18-0-18	200
65	21-0-0	200
80	34-3-8	100
90	5-10-30	150

Sprigged Zoysiagrass Fairways
ASSUME EARLY SUMMER SPRIGGING

Days After Sprigging	Suggested Ratio	Rate=lb./acre
Preplant by soil test is incorporated into 1"@ final floating	Starter *Slowly Available* N	65 lb. P 50 lb. N
15	21-0-0	160
24	15-3-15+1%Fe, 1%Mn, 1%Mg	250
31	21-0-0	160
38	15-3-15+1%Fe, 1%Mn, 1%Mg	250
52	21-0-0	200
59	15-3-15+1%Fe, 1%Mn, 1%Mg	250
69	21-0-0	160
80	15-3-15+1%Fe, 1%Mn, 1%Mg	250
90	5-10-30	150

Grow-in philosophy is basically the same as for bermudagrass but application rates are less. Because of the slower growth rate of zoysiagrass, some grow-in programming may be needed the following summer. This would be especially true if sprigging did not occur until mid- to late summer.

Bermudagrass Tees / Fairways

GRANULAR/FERTIGATION COMBINATION
ASSUME PIEDMONT/MID-SOUTH,
EARLY SUMMER SPRIGGING

Days After Sprigging	Suggested Ratio	Rate=lb./1,000 sf
Preplant by soil test is incorporated into 1"@ final floating	Starter	65 lb. P
	Slowly Available N	50 lb. N
	Soluble	
5	30-0-0	.20
7	8-12-12	.20
9	30-0-0	.20
11	8-12-12	.20
13	30-0-0	.20
15	8-12-12	.20

*At this point begin fertigation every third day for 40 days—same rate and material rotation.

GRANULAR SUPPLEMENT TO ABOVE
FERTIGATION PROGRAM

28	18-0-18, minors	225
42	21-0-0	225
60	19-19-19	200
70	5-10-30	150

If Sprigging Late Summer:
Same basic program as weather allows; however, potassium must be kept at higher ratio to better harden still immature grass off for winter.

Hybrid Bermudagrass Row Planted in Common

ASSUME EARLY SUMMER ROW PLANTING

Days After Sprigging	Suggested Ratio	Rate=lb./acre
Preplant by Soil Test	Starter	65 lb. P
5	21-0-0	225
12	15-5-15, minors	250
19	34-3-8	130
26	21-0-0	200
35	34-3-8	130
45	34-3-8	130
60	19-19-19	200
80	5-10-30	150

When Tifway is row planted into common bermudagrass fairways, it should be pushed with fertilizer initially the same as new sprigging. A nonselective postemerge herbicide prior to row planting should always be applied.

The following summer and sometimes two additional summers are needed for complete coverage of the Tifway bermudagrass. Therefore, an accelerated fertility will be needed for a more aggressive grow-in completion.

Sod Fertilization Chart

Days After Sodding	Suggested Ratio	Rate=lb./1,000 sf
Preplant by soil test is incorporated into 1"@ final floating	Starter	1 lb. P
7	23-4-10	*see chart
24	21-0-0	below—rate
38	23-4-10	by cultivar
60	18-0-18	

*FERTILITY RATE BY CULTIVAR—SOD

	Bentgrass	Ryegrass/ Bluegrass	Bermudagrass	Zoysia
7	3	4	5	3
24	3	3	5.5	3
38	4	4	4	3
60	4	5	5	5

Bentgrass Greens—High Sand

FIRST CALENDAR YEAR FERTILIZATION AFTER SEEDING; ASSUME LATE SUMMER/EARLY FALL SEEDING PREVIOUS YEAR

Estimated Date	Suggested Ratio	Primary Nutrient Rate=lb./1,000 sf
Preplant and initial grow-in outlined previous		
Mid-Feb	15-0-30	1
March 5	13-2-13	1
March 25	23-4-10	1
March 28	Micro pkg	Recom. rate
April 1	12-24-14	.5
April 15	15-0-30	.5
May 1	18-0-18	1
May 15	15-0-30	.25
June 1	0-0-28 plus Fe	.5
July 1	0-0-50	.5
August 1	0-0-50	.5
Sept 1	15-0-30	.5
Sept 15	19-19-19	1
Oct 1	18-0-18	1
Oct 7	Micro pkg	Recom. rate
Oct 15	19-19-19	1
Nov 1	18-0-18	.5
Nov 15	15-0-30	.5
Dec 15	18-0-18	1
Totals	N=10.25 lb. P=3.25 lb. K=14.5 lb.	

* These dates and rates are outline suggestions only and may vary according to weather conditions, soils, and amount of cover achieved the previous fall. Rates assuming upper transition zone climate.

    # Liquid supplements are applied on an as-needed basis dur-

ing the summer months with the granular applications of potassium.

Liquid iron(Fe) and/or micronutrient sprays can be supplemented as needed for color.

The key to remember is the accelerated percolation rate the first year and the corresponding leaching tendency. It is important to closely control water management. 10-14 day intervals with fertility has kept a balanced growth rate and allowed lower application rates to further reduce leaching.

These fertility levels outlined for the first year after planting reduce an average of about 25-30% in the second year due to establishment and maturity of the bentgrass. The infiltration rate also slows down after the first year.

APPENDIX 16

Tissue Nutrient Ranges by Turf Type

Target Ranges

	N	P	K	Ca	Mg	S
			Percent			
Creeping Bentgrass	4.50–6.00	0.30–0.60	2.20–2.60	0.50–7.50	0.25–0.30	——
Hybrid Bermudagrass						
Green/Tees	4.00–6.00	0.25–0.60	1.50–4.00	0.50–1.00	0.13–0.40	0.20–0.50
Fairways	2.30–5.00	0.15–0.50	1.00–4.00	0.35–1.00	0.13–0.50	0.15–0.50
Perennial Ryegrass	3.34–5.10	0.35–0.55	2.00–3.42	0.25–0.51	0.16–0.32	0.27–0.56
*Kentucky Bluegrass	2.51–5.40	0.27–0.49	1.73–3.08	0.27–0.58	0.13–0.32	0.18–0.24
*Zoysiagrass	1.89–2.40	0.18–0.26	1.12–1.46	0.29–0.52	0.13–0.15	0.29–0.32

	Fe	Mn	B	Cu	Zn
			ppm		
Creeping Bentgrass	100–300	50–100	8–20	8–30	25–75
Hybrid Bermudagrass					
Green/Tees	50–500	25–300	6–30	5–50	20–250
Fairways	50–500	25–300	6–30	5–50	20–250
Perennial Ryegrass	97–934	30–73	5–17	6–38	14–64
*Kentucky Bluegrass	102–189	18–48	6–9	8–33	19–88
*Zoysiagrass	161–273	26–31	6–12	1–17	35–55

*Survey results

Appendix 17

First Year Budget Line Items/Areas to Address

The first year budget of any new course will have additional, specific expenses that are unique to a "maturing" golf course. Problems such as drainage, birdbaths, and unexpected tree removal are prime examples.

During the first full year's time frame; that, is the first year budget after grow-in, the golf course matures and things happen that in general will not be dealt with again except on a limited basis. Therefore, these problem areas must be projected and budgeted for, to get all "out-of-normal" special projects completed.

The most predominant ones include:

1. Additional labor—to get the special projects completed without sacrificing routine maintenance needs.
2. Tree removal—some trees die unexpectedly after construction despite the best protection efforts. Don't forget stump removal. Also correct the shade/air movement problems not identified in construction/grow-in.
3. Tree replacement—should be in accordance with the Master Plan— coordinate with architect. This is planned for here instead of during grow-in because you can better evaluate the need for replacements, deletions, or additions.
4. Birdbaths—settling in fairways, tees, collars—sand and additional labor is needed. Continual topdressing can solve most problems which is a lot less expensive. Don't use sand that is coarse. Mason

sand is the best type gradation, preferably a river sand with a small silt and clay component.

5. Additional drainage—sand, gravel, pipe, outlets— suggested budget is about twelve 100-ft. runs with outlets.

6. Additional sod—estimate 20,000 sq. ft.

7. Additional slicing/aeration needed over entire course—very successful with eliminating construction scars and final smoothing of playing surfaces, slopes, and immediate roughs. Also great at increasing turf density.

8. Additional fertilizer—average about 20% above normal maintenance needs to finish establishment in difficult areas.

9. Additional irrigation heads needed—irrigation contractor can give you a unit cost addition— estimate 10-15 needed.

10. Additional repairs—the course is still a little rough in areas and a few more breakdowns will occur despite new equipment. Oftentimes a reduced M&R budget is used the first year but a regular M&R line amount is recommended.

Providing owner/developers with detailed information and projections through grow-in and the first year shows tremendous knowledge and organization on your part. Be ahead of the curve during a usually difficult funding time by making owners aware of the costs. When golf course needs are projected, these budgets, that usually are cut short, become funded to meet the needs.

Appendix 18

Correct French Drain Profile

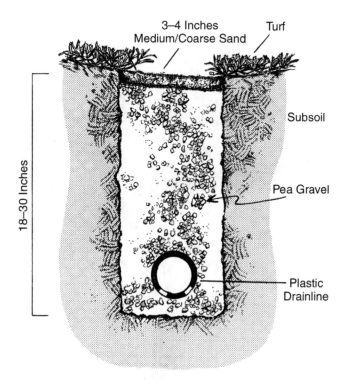

3–4 Inches
Medium/Coarse Sand

Turf

Subsoil

18–30 Inches

Pea Gravel

Plastic
Drainline

APPENDIX 19

Deep Root Tree Fertilization Injection Pattern

Drip Line

3 Feet

3 Feet

APPENDIX 20

Tree Placement and Shape Affect
Playability and Turf Health

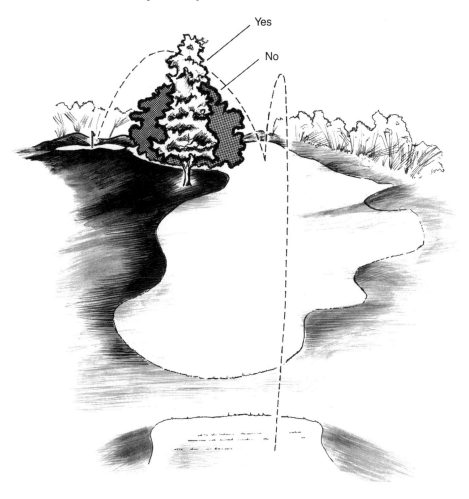

GLOSSARY

Aerobic: Oxygen present

Anaerobic: No oxygen present

Asbuilts: Corrected, architectural/engineering drawings completed after construction to show changes, structures, locations, and other details

Base Saturation: Measure of a soil's cation ratios

Birdbaths: Depressions that develop from settling or washing and hold water

BMPs: Best Management Practices—Guidelines for environmentally responsible turf management

Budget, Grow-In: Budget required for the time period from planting until opening

Bulkheads: Retainer wall used along water or where sloped banks are not feasible

Bunker Liners: Geotextile fabric most commonly used to prevent migration of stones upward into the bunker sand

Bury Pits: Pits used to bury debris during clearing—stumps, remains from burn piles, soil strippings which contain roots and organic matter

CEC: Measure of a soil's cation holding capacity

Choker Layer: Coarse sand layer in a USGA spec green

Critical Path: Order of events that must occur during any selected phase of construction/grow-in/development

Cultural Practices: Management practices that include aerification, vertical mowing, slicing, topdressing, dragging, or brushing

Deep Roughs: Roughs beyond immediate roughs (adjacent fairways) and which commonly contain natural or wildlife management areas

Dormant Seeding: Seeding done during the winter in preparation for next spring. Winter rains move seed into better soil contact

Dormant Sodding: Sodding done during the winter to protect a newly shaped slope and reduce erosion

Erosion and Sediment Control: Plans, management programs, and structures used to prevent erosion damage and siltation

Fertigation: Liquid fertilizer injection/application through the irrigation system

Floating: Final smoothing of the surface or seedbed preparation. All grades to outlets and depressions are finalized

Flushing: An outlet (stub-out) in the drainline (usually the highest end) where high pressure flushing can be accomplished

Forced Air: Air blown into the soil profile via the drainage system by high volume pump. Reversal can assist in pulling water out of the root zone

Grow-in Layer: A very concentrated mat or layer of thatch created on high-sand greens during grow-in when the growth rate is very high for establishment. This organic layer can be a management problem if not removed after grow-in is completed

Haul Roads: Access roads over the course for construction traffic

Hydroseeding: A seed/fertilizer and mulch mixture in a water suspension sprayed onto an area via high volume pump

IPM: Integrated Pest Management utilizes reduced chemical and water inputs into turf management practices for greater environmental response

Long-Range Plans: Management plans for future development/planning to assist in budgeting or prioritizing

Master Plan: The complete development plan, usually including golf course, club amenities, and housing development

Microbial Activity: Soil microbiological functions and activity

Microinjection: Soil conditioner, micronutrients, or fertilizer injected in small doses on a continual basis during the growing season via the irrigation system (see Fertigation)

Natural Areas: Wildlife habitat areas, wildflowers or native species areas incorporated in deep rough (out-of-play areas) for reduced maintenance costs, improved aesthetics, and wildlife/Audubon promotion

Non-Point Source Pollution: Contamination of an off-site area from fertilizer or chemical runoff or leaching

Owner Rep: Golf course superintendent serving as the on-site management or official contact person to contractors during construction

Punch List: List of to-be-completed items developed at end of construction project by owner rep, contractor, architect

Quality Control: On-site management of construction materials, techniques, and progress to ensure a quality product

RFPs: Request for payment—itemized statements

Roguing: Routine walk inspections and hand-removing contamination (i.e., *Poa annua* spots) in bentgrass greens or fairway bermudagrass spots in putting green bermudagrass

Rowplanting: Planting bermudagrass or zoysiagrass sprigs into existing turf cover areas—done in 6" or 12" centers

Scouting: Periodically inspecting the course for pest activity, irrigation cycling, etc. and mapping/recording observations

Slicing: Cultural management practice of vertical knife aeration to smooth and firm the soil and promote lateral growth of turf

Staging Areas: Preselected area or areas on site where materials are delivered, stockpiled for blending, and redistributed onto the course

Tree Protection: Protecting tree root systems during construction from grading and traffic

Walk-Throughs: Periodic site inspections of construction progress by owner rep, golf course architect, contractor

Washout Repair: Erosion damage filling and establishment

Wet Springs: Subsurface water that now breaks the surface in a new cut area during the rainy season when water table rises

Wicking Barrier: Vertical plastic barrier placed around sand profile core of green to prevent lateral movement of water for sand to existing soil perimeter. Also reduces root migration of trees

REFERENCES

1. Balogh, J.C. and W.J. Walker. *Golf Course Management & Construction*, Lewis Publishers, Chelsea, MI. 1992.
2. Beard, J.B. *Turf Management for Golf Courses*, Burgess Publishing Company, Minneapolis, MN. 1982.
3. Burpee, L.L. *A Guide to Integrated Control of Turfgrass Diseases*, Volume I, GCSAA Press, Lawrence, KS. 1993.
4. Christians, N. *Fundamentals of Turfgrass Management*, Ann Arbor Press, Chelsea, MI. 1998.
5. Daniel, W. H. and R. P. Freeborg. *Turf Manager's Handbook*, Harvest Publishing Company, Cleveland, OH. 1987.
6. Georgia Department of Natural Resources, *Erosion & Sediment Control Manual*, 1998.
7. *Golf Course Management Magazine*. Golf Course Superintendents Association of America.
8. Harker, D., S. Evans, and K. Harker. *Landscape Restoration Handbook*, Lewis Publishers, Chelsea, MI. 1993.
9. Jones, J.B., Jr. *Plant Nutrition Manual*, CRC Press, Boca Raton, FL. 1998.
10. Kent, D. *Applied Wetland Science and Technology*, Lewis Publishers, Chelsea, MI. 1994.
11. Kinsey, N. and C. Walters. *Hands-on Agronomy*, Acres USA, Metairie, LA. 1995.
12. *Landscape Management Magazine*. Avanstar Communications, Inc.
13. Leslie, A.R. *Handbook of Integrated Pest Management for Turf and Ornamentals*, Lewis Publishers, Chelsea, MI. 1994.
14. MacCarthy, P., C.E. Clapp, R.L. Malcolm, and P.R. Bloom.

Humic Substances in Soil and Crop Sciences: Selected Readings, 1990.

15. Madison, J.H. *Agronomy Journal* 58:441–443, 1966.
16. Mills, H.A. and J.B. Jones, Jr. *Plant Analysis Handbook II,* MicroMacro Publishing, Athens, GA. 1996.
17. Schuman, G.L., P.J. Vittum, M.L. Elliott, and P.P. Cobb. *IPM Handbook for Golf Courses,* Ann Arbor Press, Clelsea, MI. 1998.
18. State Soil Water Conservation Commission. *Manual for Erosion and Sediment Control in Georgia.*
19. *Turfgrass Selection Guide.* Loft's Seed, 3rd edition.
20. *TurfGrass TRENDS.* Avanstar Communications, Inc.
21. UGA Cooperative Extension Service. Weeds of Southern Turfgrasses. Georgia College of Agriculture, Athens, GA.
22. USGA. *USGA Green Section Record.* March/April 1993.
23. USGA. *Wastewater Reuse for Golf Course Irrigation,* Lewis Publishers, Chelsea, MI. 1994.
24. Uva, R.H., J.C. Neal, and J.M. DiTomaso. *Weeds of the Northeast,* Comstock Publishing Associates, Ithaca, NY. 1997.
25. Washington State Department of Development and Environment Services. *Best Management Practices for Golf Course Development and Operations.*
26. *Western Fertilizer Handbook-Horticulture Edition.* California Fertilizer Association. Interstate Publishers, Inc., Danville, IL. 1990.

Specific Articles/Publications

27. Cooper, R.J. Soil Microbes and Their Effect on Turfgrass Health. *North Carolina Turfgrass,* June/July 1997.
28. Dernoeden, P.H. Fine Fescues on Golf Courses. *Golf Course Management,* April 1998.
29. Harivandi, M. Ali and J.B. Beard. How to Interpret a Water Test Report. *Golf Course Management,* June 1998.

30. Kupier, M. Have a Plan. *Golf Course Management,* August 1997.

31. McLaughlin, R.M. Turfgrass Water Quality Issues, Challenges and Misconceptions. *North Carolina Turfgrass,* October/November 1998.

32. Mikkelsen, L. Taming Wild Waters. *USGA Green Section Record,* March/April 1997.

33. Miller, G.A. Using Reverse Osmosis to Make Irrigation Water. *USGA Green Section Record,* November/December 1998.

34. Stairs, N. Methods to Stop Moving Soil. *Landscape Management,* July 1998.

35. *The Guide to Estimating Cost for Golf Course Construction,* GCBAA. 1996.

36. Watschke, T. L. Control Weeds in Newly Established Turf. *Grounds Maintenance,* August 1995.

37. White, B. If You Care for Microbes, Microbes Will Care for Your Turf. *Golf Course Management,* September 1998.

38. White, B. What To Do After the Contractor Leaves. *Golf & sportsTURF,* April 1991.

Association Offices for Further Information

American Society of Golf Course Architects (ASGCA). Chicago, IL.
Audubon International. Selkirk, NY.
Golf Course Builders Association of America (GCBAA). Chapel Hill, NC.
USGA Green Section. Far Hills, NJ.

INDEX

Erosion & Sediment Control Plan Legend
Erosion & Sediment Control Plan Excerpt

Georgia
Uniform Coding Form
for Soil Erosion and Sediment Control Practices

Code	Practice	Detail	Map Symbol
(Cd)	Checkdams		
(Ch)	Channel Stabilization	Stone-Lines Waterways / Gravel Bedding	
(Co)	Construction Exit		(Label)
(Rp)	Riprap		

Riprap

(Sd1)	Sediment Barrier		Type
(Sd3)	Sediment Basin Temporary		
(Sd2)	Sediment Trap Temporary		
(Sr)	Crossing Temporary		
(St)	Storm Drain Inlet/Outlet Protection		(Label)

Vegetative Practices

BF	Buffer Zone		BF (Label)
Ds1	Disturbed Area Stabilization (with Mulching Only)		Ds1
Ds2	Disturbed Area Stabilization (With Temporary Seeding)		Ds2
Ds3	Disturbed Area Stabilization (With Permanent Vegetation)		Ds3
Du	Dust Control On Disturbed Areas		Du

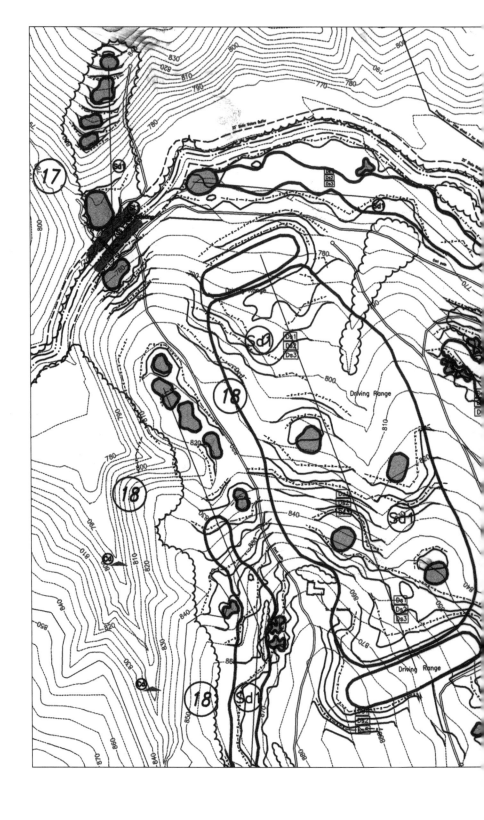